Lectures on Analytical Mechanics

Lectures on Analytical Mechanics

G. L. KOTKIN
V. G. SERBO
A. I. CHERNYKH

Novosibirsk State University, Russia

Translation from the third enlarged Russian edition by O.V. Karpushina and V.G. Serbo

OXFORD
UNIVERSITY PRESS

OXFORD
UNIVERSITY PRESS

Great Clarendon Street, Oxford, OX2 6DP,
United Kingdom

Oxford University Press is a department of the University of Oxford.
It furthers the University's objective of excellence in research, scholarship,
and education by publishing worldwide. Oxford is a registered trade mark of
Oxford University Press in the UK and in certain other countries

Published in the United States of America by Oxford University Press
198 Madison Avenue, New York, NY 10016, United States of America

British Library Cataloguing in Publication Data
Data available

Library of Congress Control Number: 2023949600

ISBN 9780198894674
ISBN 9780198894681 (pbk.)

DOI: 10.1093/oso/9780198894674.001.0001

Printed and bound by
CPI Group (UK) Ltd, Croydon, CR0 4YY

Preface

It is clear that analytical mechanics is necessary, one way or another, for a physicist specializing in mechanics (hydrodynamics, gas dynamics, etc.). But what about students pursuing careers in plasma physics, nuclear physics, quantum optics, radio physics, and more? Indeed it is! The fact is that analytical mechanics is the introductory chapter to theoretical physics; the methods and ideas developed in there are vitally important to all other branches of theoretical physics. Lagrangian and Hamiltonian formalisms, normal oscillations, adiabatic invariants, Liouville's theorem, canonical transformations – all these concepts are the basics, without the knowledge of which a more in-depth study of field theory, statistical physics, and quantum mechanics is impossible. In any serious book on physics, a standard phrase: *The Hamiltonian in this case has the form* ... appears almost unavoidably (see, for example, *Physicists Keep Joking*. S. 154).

This book was written by working physicists for the students at physics faculties of universities. It is based on our many years of experience in lecturing and conducting seminars at the Physics Department of Novosibirsk State University in Russia. These lectures were delivered once a week during the fourth semester, alongside the second course of electrodynamics and before the course on quantum mechanics.

In this edition, the key notations used are as follows:

m, e, $\mathbf{r}$, $\mathbf{p}$ and $\mathbf{M} = [\mathbf{r}, \mathbf{p}]$ – mass, charge, radius vector, momentum, and angular momentum of a particle, respectively;

L, H, E and U – Lagrangian function, Hamiltonian function, energy, and potential energy of a system, respectively;

$\mathbf{E} = -\nabla\varphi - \frac{1}{c}\frac{\partial \mathbf{A}}{\partial t}$ and $\mathbf{B} = [\nabla, \mathbf{A}]$ – electric and magnetic field intensities, respectively;

φ and $\mathbf{A}$ – scalar and vector potentials, respectively, of the electromagnetic field;

c – velocity of light;

$d\Omega$ – solid angle element; and

e_{ijk} – the completely antisymmetric tensor of the third rank, $e_{123} = e_{231} = e_{312} = 1$, $e_{132} = e_{321} = e_{213} = -1$, the remaining components of e_{ijk} are zero.

For problems related to particle motion in electromagnetic fields, Gaussian units are used, while SI units are used for problems related to electrical circuits.

This course has the following characteristics:

(i) The main feature is the gradual introduction of increasingly complex aspects of analytical mechanics, to actively engage students. The course begins with a review of previously taught material, including Newton's equations, motion in a central field, and scattering problems. Next, the Lagrangian equations are derived from the Hamilton principle, and their validity is confirmed by reducing them to Newton's equations. This approach facilitates the learning of new concepts in Lagrangian mechanics. The course then covers linear and nonlinear oscillations, Hamiltonian formalism, and rigid-body motion, which are traditionally established chapters of analytical mechanics taught as part of the theoretical physics course. However, more advanced topics such as the general theory of dynamic equations, improved perturbation theory for non-linear oscillations, and dynamic chaos are beyond the scope of this course and should be covered in additional courses.

(ii) Whenever possible, we draw analogies and make comparisons to electrodynamics, quantum mechanics, and statistical mechanics, as we cannot resist the temptation to do so.

(iii) The book includes most of the problems discussed in our seminars. These problems are sourced from *The Collection of Problems in Classical Mechanics* by Kotkin and Serbo [3], which serves as a guidebook for the course, where students and teachers could find not only the problems themselves, but also their solutions.

(iv) This new extended edition includes ten additional sections aimed at illustrating the theoretical material. For instructors, these sections can serve as the foundation for additional courses. Meanwhile, students seeking to broaden their understanding of various applications of analytical mechanics can use these sections as supplementary reading material or for self-study.

(v) Thoughtful students will find supplements that expand on and clarify the main text.

Formulae are numbered using two digits. For example, (3.7) signifies formula (7) from § 3. References to formulae from the same paragraph are abbreviated without specifying the paragraph number.

Finally, it is also worth mentioning that a new electronic textbook *Selected Chapters of Analytical Mechanics* (*Electronic Textbook with Dynamical Interactive Illustrations*) by Serbo and Cherkassky [11] has recently been released. This textbook offers some computer-illustrated problems of analytical mechanics, including real-time particle motion in various fields and under different initial conditions. Moreover, the reader has the ability to independently manipulate and experiment with the problem parameters, facilitating a clearer understanding of the underlying concepts rather than relying on verbose descriptions.

In conclusion, we will point out that there are many good books on analytical mechanics, starting with Landau and Lifshitz's *Mechanics* (first volume course of theoretical physics) [1] and Goldstein, Poole, and Safko's *Classical mechanics* [2]. Among the more recent publications, we suggest *Classical Dynamics* by D. Tong (available online at http://www.damtp.cam.ac.uk/user/tong/dynamics/), which serves as the coursebook for *Mechanics* at Cambridge University.

This edition was prepared by us following the untimely death of our dear friend and co-author, Gleb Leonidovich Kotkin (1934–2020).

V. G. Serbo; email: vgserbo@gmail.com
A. I. Chernykh; email: chernykh@iae.nsk.su

Contents

I Newton's mechanics. Central field. Scattering 1

§ 1 One-dimensional motion in a potential field. Period of oscillations 1

Problems 3

§ 2 Motion in a central field 3

2.1 Radial motion 4

2.2 Orbits of the motion 5

Problem 7

§ 3 Kepler's problem 7

3.1 Orbits 7

3.2 Elliptical orbit. Kepler's laws 9

3.3 The Laplace vector as the additional integral of motion in the Kepler problem 11

Problem 12

§ 4 Perihelion precession under the perturbation $\delta U(r)$ 12

Problem 14

§ 5 Motion in the central field $U(r) = -\frac{\alpha}{r} + \frac{\beta}{r^2}$ 15

§ 6 Isotropic oscillator 17

§ 7 The two-body problem 19

§ 8 Scattering cross section. Rutherford's formula 20

8.1 Setup of the scattering problem 20

8.2 Small angle scattering 21

8.3 Rutherford's formula 24

Problems 26

§ 9 Virial theorem 26

Problem 28

II Lagrangian mechanics 29

§ 10 Lagrangian equations 29

10.1 Lagrangian equations for the non-relativistic particles in a potential field as a covariant notation of Newton's equations 29

10.2 Generalized coordinates and momenta 30

Problem 31

§ 11 Principle of a least action 31

11.1 Hamiltonian principle. Covariance of the Lagrangian equations with respect to replacement of coordinates 31

11.2 Transformation of the Lagrangian function under transformation of coordinates and time 33

§ 12 Lagrangian function for a particle in an electromagnetic field. Ambiguity in the choice of the Lagrangian function 35

§ 13 Classic Zeeman effect 36
13.1 Charged particle in the Coulomb and magnetic fields 36
13.2 Strong magnetic field. Drift 38
§ 14 Lagrangian function in the relativistic case 40
§ 15 The Lagrangian function for systems with ideal holonomic constraints 42
Problems 45
§ 16 Cyclic coordinates. Energy in the Lagrangian approach 46
16.1 Cyclic coordinates 46
16.2 Energy in the Lagrangian approach 46
16.3 Is the energy in the Lagrangian approach equal to the sum of kinetic and potential energies? 48
16.4 Ambiguity in the definition of energy 49
§ 17 Symmetry and integrals of motion. Noether's theorem 50
17.1 Examples 50
17.2 Generalization 51
17.3 Noether's theorem 51
Problems 53
§ 18 Fundamental conservation laws for a closed system of particles 54
§ 19 Galilean transforms 55
§ 20 Non-inertial frames of reference 57
20.1 Translational reference frame 57
20.2 Rotating reference frame 58
20.3 Larmor's theorem 59
§ 21 Deviation of a freely falling body from the vertical 60
§ 22 Effective Lagrangian function for electromechanical systems 62

III Oscillations 65

§ 23 Linear oscillations 65
23.1 One degree of freedom 65
23.2 Oscillations of systems with many degrees of freedom 66
23.3 Flat double pendulum 68
§ 24 Orthogonality of normal oscillations. The case of frequency degeneracy 70
24.1 Orthogonality of normal oscillations 70
24.2 The case of frequency degeneracy. Normal coordinates 71
24.3 Oscillations of weakly coupled systems. Beats 72
Problems 75
§ 25 Forced oscillations. Resonances 75
Problem 78
§ 26 Oscillations in the presence of a friction force 79
§ 27 Oscillations in the presence of gyroscopic forces 82
27.1 Gyroscopic forces 82
27.2 Small oscillations of a charged particle in a magnetic field 83
27.3 Oscillator in a uniform magnetic field 86
27.4 Anti-oscillator in a uniform magnetic field 88
27.5 Penning trap 89

27.6 Particle inside a smooth rotating paraboloid in the field of gravity 91
27.7 Lagrange points in the solar system 92
§ 28 Oscillations of symmetric systems 93
Problems 96
§ 29 Oscillations of molecules 96
Problems 98
§ 30 Oscillations of linear chains 98
30.1 Equations of motion and boundary conditions 98
30.2 Travelling waves 99
30.3 Standing waves and spectrum 102
Problems 104
§ 31 The Born chain. Acoustic and optical oscillations of linear chains 104
§ 32 Forced oscillations of linear chains under the action of a harmonic force 107
§ 33 Non-linear oscillations. Anharmonic corrections 108
33.1 One-dimensional non-linear oscillations 109
33.2 Multidimensional non-linear oscillations. Combination frequencies 111
§ 34 Non-linear resonances 113
§ 35 Classical model of the Fermi resonance in a CO_2 molecule 117
§ 36 Parametric resonance 119
§ 37 Motion in a rapidly oscillating field 122
Problem 123

IV Hamiltonian mechanics 124

§ 38 Hamiltonian equations 124
38.1 Hamiltonian function. Hamiltonian equations 124
38.2 Integrals of motion in the Hamiltonian approach 127
Problems 130
§ 39 Variational principle for the Hamiltonian equations 130
§ 40 Poisson brackets 131
40.1 Definition and main properties 132
40.2 Jacobi identity and Poisson's theorem 134
Problem 136
§ 41 Dynamic symmetry of the Kepler problem 136
§ 42 Classical model of EPR and NMR 137
42.1 Equations of motion of the vector $\mathbf{M}(t)$ 138
42.2 Motion of the vector $\mathbf{M}(t)$ in a rotating magnetic field 138
§ 43 Canonical transformations 140
43.1 Definition of the canonical transformation. Generating function 140
43.2 Other generating functions 143
Problems 145
§ 44 Canonical transformations and Poisson brackets 145
44.1 Invariance of the Poisson brackets with respect to the canonical transformations 145

44.2 Necessary and sufficient criterion that the transformation is canonical 146
§ 45 Examples of canonical transformations 147
Problem 149
§ 46 Action along the true trajectory as a function of initial and final coordinates and time 149
46.1 Properties of $S(q, t)$ 150
46.2 Motion of a system as a canonical transformation 151
46.3 Proof of the Noether's theorem 151
§ 47 The Liouville theorem 152
47.1 Invariance of a phase volume with respect to canonical transformations 152
47.2 Focusing lens 154
Problem 155
§ 48 The Hamilton–Jacobi equation 155
48.1 The Hamilton–Jacobi equation. Separation of the variables 155
48.2 Motion of a relativistic particle in the field $U(r) = -\frac{\alpha}{r}$ 157
48.3 Opto-mechanical analogy 159
Problem 160
§ 49 Angle and action variables 160
49.1 Systems with one degree of freedom 160
49.2 Systems with many degrees of freedom 164
49.3 Hamiltonian function explicitly depending on time 164
§ 50 Adiabatic invariants 166
50.1 Setup of the problem and result 166
50.2 Adiabatic invariant for a particle in a box 168
50.3 Conservation of the adiabatic invariant 168
Problems 170
§ 51 Motion of a system with many degrees of freedom. Dynamic chaos 170

V Rigid-body motion 174

§ 52 Kinematics of a rigid body 174
§ 53 Momentum, angular momentum and kinetic energy of a rigid body 176
53.1 Momentum of a rigid body 176
53.2 Angular moment of a rigid body 177
53.3 Kinetic energy of a rigid body 178
53.4 Inertia tensor of a rigid body 179
Problems 181
§ 54 Equations of motion of a rigid body. Examples 182
54.1 Equations of motion of a rigid body. Euler equations 182
54.2 Free motion of spherical and symmetrical tops 183
54.3 Fast top in the field of gravity 186
Problems 187
§ 55 Effect of tidal forces on the length of a day and month 188
§ 56 Euler angles 189

Supplements 192

A Elements of the calculus of variations 192

B Systems with constraints 194

B.1 Systems with ideal holonomic constraints 194

B.2 Reaction forces of constraints 197

B.3 Indefinite Lagrangian multipliers. Ideal non-holonomic constraints 198

C Hill equation, Mathieu equation, and parametric resonance 200

C.1 General properties of the Hill equation 200

C.2 Mathieu equation 201

C.3 Parametric resonance on the fundamental harmonic $\gamma = 2\omega_0$ 202

C.4 Parametric resonance at $\gamma = \omega_0$ 204

D Generalization of canonical transformations 206

D.1 Time and energy as canonical variables 206

D.2 Canonical transformations involving time and energy 207

E Differential forms and canonical transformations 208

E.1 Differential forms 208

E.2 New definition of canonical transformations 211

E.3 Conservation of the phase volume under canonical transformations 211

E.4 Invariance of the Poisson brackets under the canonical transformations 211

Bibliography 213

Index 214

CHAPTER I

Newton's mechanics. Central field. Scattering

In what follows, it is assumed that such concepts as inertial frame of reference, material point (particle), mass, force, potential energy, and Newton's laws are well known from the course of general physics. In particular, in celestial mechanics, the frame of reference, which is at rest with respect to the centre of mass of the solar system and whose three mutually orthogonal coordinate axes are fixed to stars, is the inertial frame with a high accuracy.

§ 1 One-dimensional motion in a potential field. Period of oscillations

Let a particle of mass m move in some inertial frame of reference along the x-axis in the potential field $U(x, t)$. As is well-known, the equation of motion (Newton's equation)

$$m\ddot{x} = F_x(x, t) = -\frac{\partial U(x, t)}{\partial x} \tag{1.1}$$

with the initial conditions $x(t_0) = x_0$ and $\dot{x}(t_0) = v_0$ has the only solution $x(t)$. If the potential energy does not depend on t, i.e. $U = U(x)$, then energy is conserved for motion in such a field,

$$E = \frac{1}{2} m\dot{x}^2 + U(x) = \text{const}\,, \tag{1.2}$$

which can be verified by direct differentiation over time

$$\frac{dE}{dt} = \left(m\ddot{x} + \frac{dU(x)}{dx} \right) \dot{x} = 0\,.$$

Energy is an example of an *integral of motion*, i.e. such a function of coordinates and velocities which is conserved for motion of a system. For one-dimensional motion, the presence of such an integral of motion allows us to use the first-order equation (2) instead of the second-order one (1) and to find the law of motion $x(t)$ in quadratures.

Lectures on Analytical Mechanics. G. L. Kotkin, V. G. Serbo, A. I. Chernykh, Oxford University Press. © G. L. Kotkin, V. G. Serbo, A. I. Chernykh (2024). DOI: 10.1093/oso/9780198894674.003.0001

Indeed, first, we find the constant

$$E = \frac{1}{2}mv_0^2 + U(x_0)\,,$$

and then, using it in the first-order equation

$$\frac{dx}{dt} = \pm\sqrt{\frac{2}{m}\,[E - U(x)]} \quad (\text{when} \quad \dot{x} \gtrless 0), \tag{1.3}$$

we separate the variables and get the answer in quadratures:

$$t = \pm\sqrt{\frac{m}{2}}\int_{x_0}^{x} \frac{dx}{\sqrt{E - U(x)}} + t_0\,. \tag{1.4}$$

The points x_i, at which $U(x_i) = E$, define the boundaries of the region of the particle motion. At these points, velocity is zero, $v_i = 0$, but the acceleration $a_i = -U'(x_i)/m$ may be non-zero.

The law of the energy conservation allows us to obtain the integral of motion (4); however, this integral cannot always be calculated analytically. Nevertheless, the qualitative behaviour of the solution can be understood even for the arbitrary potential energy $U(x)$. Let us consider an example of the potential energy with a local maximum U_m (Fig. 1). Since the kinetic energy $T = E - U(x) > 0$, then at the energy $E < U_m$, the motion is possible only in these two regions: in the interval $x_1 \le x \le x_2$ and on the half-line $x \ge x_3$. Accelerations $a_{1,2,3} \ne 0$; moreover, $a_{1,3} > 0$, while $a_2 < 0$, so near the points $x_{1,2,3}$ the motion turns out to be approximately uniformly accelerated. Therefore, the points x_i are *the turning points*. In the region $x_1 \le x \le x_2$, the particle oscillates with a period

$$T = \sqrt{2m}\int_{x_1}^{x_2} \frac{dx}{\sqrt{E - U(x)}}; \tag{1.5}$$

its motion is *finite*. In the region $x \ge x_3$, the particle goes to infinity; its motion *is infinite*.

The special case is the motion of a particle with energy equal to local maximum, $E = U_m$. In this case, at the point of the local maximum x_m, the force $F_x(x_m) = -U'(x_m)/m = 0$, i.e. not only the velocity, but also the acceleration of the particle vanishes. Let us consider

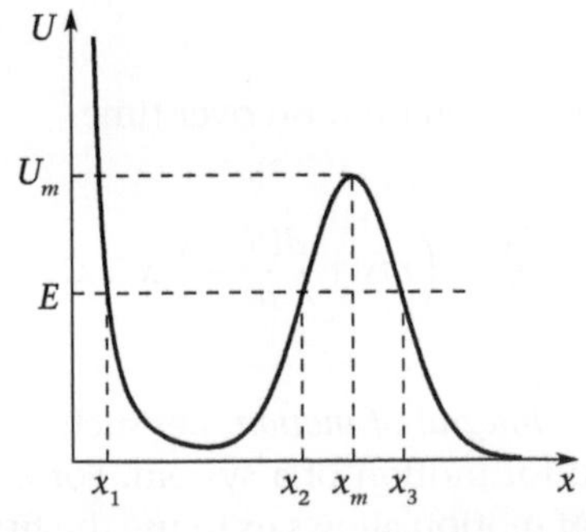

Figure 1 Boundaries of motion in a given potential field

the following example when $U'(x_m) = 0$, but $U''(x_m) \neq 0$. In this case, the potential energy near points x_m has the form

$$U(x) = U_m + \frac{1}{2} U''(x_m)\,(x - x_m)^2 + \dots, \quad U''(x_m) < 0$$

and the solution of equation (4), corresponding to the motion near point x_m, reads

$$x(t) = x_m - (x_m - x_0)\,\mathrm{e}^{-\lambda(t-t_0)}, \quad \lambda = \sqrt{-\frac{U''(x_m)}{m}}.$$

That is, $x(t) \to x_m$ only for $t \to \infty$. Thus, in this case, the particle's approach to the point x_m from the left or from the right will occur indefinitely. Therefore, the period of oscillations (5) goes to infinity when energy $E \to U_m$.

Problems

1.1. Describe qualitatively the motion of a particle of mass m in the potential field (for $x > 0$)

$$U(x) = V\frac{a^2}{x^2}\left(1 - \frac{a}{x}\right)^2$$

at different values of the particle energy.

1.2. Describe the motion of a particle in the field $U(x) = -Ax^4$ for the case when its energy is zero. At the initial moment, the particle is at point $x(0) = a$. Consider the cases $\dot{x}(0) > 0$ and $\dot{x}(0) < 0$.

The same question for the field $U(x) = -kx^2$.

1.3. Upon observing the finite motion of a particle over a time which is much longer compared to the period of the particle's motion T, we can introduce the probability distribution of detecting a particle near the point x. More precisely, the probability dw to find a particle in the interval $(x, x + dx)$ can be defined as the ratio of the entire time $2\,dt$ spent by the particle on this interval during the period of the particle motion T to this period: $dw = 2\,dt/T$. Find the probability density $dw(x)/dx$ for particle motion in the field of oscillator $U(x) = \frac{1}{2}\,m\omega^2 x^2$ with amplitude a.

§ 2 Motion in a central field

The potential energy of the *central field* depends only on the modulus of the radius vector $U(\mathbf{r}) \equiv U(r)$. Two important examples of such motion are that of a planet in the solar system and that of an electron in a hydrogen atom. In addition, the motion in the central field is one of a few examples of the problems solved in quadratures for an arbitrary dependence $U(r)$. Of course, the equations of such a motion can be solved numerically, but, if calculated for an extended period of time and for complex conditions, this becomes a task exceeding the capabilities of even powerful computers (see the example of the elastic scattering of balls in § 51).

A large number of educational problems can be found in [3] § 2. We would suggest the simple and exactly solvable problems 2.3 and 2.5 and an interesting problem 2.40, which is suitable for more advanced readers.

2.1 *Radial motion*

When a particle moves in a central field, the force acting on the particle[1]

$$\mathbf{F} = -\frac{\partial U}{\partial \mathbf{r}} = -\frac{dU}{dr}\frac{\mathbf{r}}{r}$$

is directed either along the radius vector or against it. In that field, not only energy

$$E = \frac{1}{2}m\mathbf{v}^2 + U(r)\,, \tag{2.1}$$

but also the angular momentum

$$\mathbf{M} = m[\mathbf{r}, \mathbf{v}] \tag{2.2}$$

remain constant, since[2]

$$\frac{d\mathbf{M}}{dt} = [\mathbf{r}, \mathbf{F}] = 0\,.$$

From (2) it follows that the orbit of the particle is in the plane perpendicular to the constant vector **M**; let it be the *xy*-plane. By introducing polar coordinates r and φ in this plane (Fig. 2), we obtain

$$E = \frac{1}{2}m\dot{r}^2 + \frac{1}{2}m(r\dot{\varphi})^2 + U(r)\,, \tag{2.3}$$

$$\mathbf{M} = (0, 0, M)\,, \quad M = mr^2\dot{\varphi}\,. \tag{2.4}$$

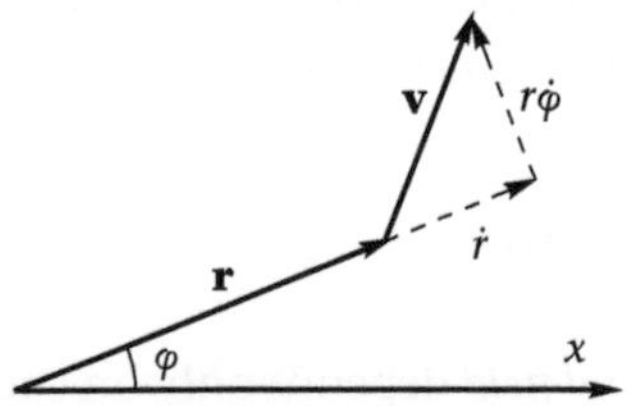

Figure 2 Components of velocity in polar coordinates r, φ

[1] Derivatives with respect to the vector are defined as follows:

$$\frac{\partial}{\partial \mathbf{r}} \equiv \left(\frac{\partial}{\partial x}, \frac{\partial}{\partial y}, \frac{\partial}{\partial z}\right).$$

[2] Note the following. It is easy to prove the fact of conservation if one knows the form of the function conserved. What is much harder, however, is to guess this form if the conservation of angular momentum is unknown.

Using (4) to eliminate $\dot{\varphi}$ from (3), we find

$$E = \frac{1}{2} m\dot{r}^2 + U_{\text{eff}}(r)\,, \tag{2.5}$$

where

$$U_{\text{eff}}(r) = U(r) + \frac{M^2}{2mr^2}\,.$$

The radial motion can thus be reduced to the one-dimensional motion in a field with the *effective potential energy* $U_{\text{eff}}(r)$, including the *centrifugal energy* $M^2/(2mr^2)$ (cf. (1.2)). From (5) we find

$$\frac{dr}{dt} = \pm\sqrt{\frac{2}{m}\,[E - U_{\text{eff}}(r)]} \quad (\text{when} \quad \dot{r} \gtrless 0)$$

or

$$dt = \pm\sqrt{\frac{m}{2}}\,\frac{dr}{\sqrt{E - U_{\text{eff}}(r)}}\,, \tag{2.6}$$

whence we obtain the dependence $t(r)$:

$$t = \pm\sqrt{\frac{m}{2}} \int_{r_0}^{r} \frac{dr}{\sqrt{E - U_{\text{eff}}(r)}} + t_0\,.$$

This equation is completely analogous to Eq. (1.4) for the one-dimensional motion. Therefore, to find $r(t)$ one can repeat all that was done for the dependence $x(t)$ in § 1 replacing the potential energy $U(x)$ with the effective potential energy $U_{\text{eff}}(r)$. Thus, the points r_i at which $U_{\text{eff}}(r_i) = E$ define the boundaries of the area of the particle motion along the radius. At these points the radial velocity $\dot{r}_i = 0$ but the radial acceleration $\ddot{r}_i = -U'_{\text{eff}}(r_i)/m$ is usually non-zero. In this case, the points r_i are the turning points for the radial motion of the particles. For example, if the dependency $U_{\text{eff}}(r)$ on r has the same form as the dependence of $U(x)$ on x in Fig. 1, then at energy $E < sU_m$, the motion is possible only in two areas: the finite motion in the interval $r_1 \le r \le r_2$ and the infinite one in the region $r \ge r_3$. For the finite motion, the period of radial oscillations is (cf. (1.5))

$$T_r = \sqrt{2m} \int_{r_1}^{r_2} \frac{dr}{\sqrt{E - U_{\text{eff}}(r)}}\,. \tag{2.7}$$

2.2 *Orbits of the motion*

To derive the equation for an orbit, we use (4) in the form

$$dt = \frac{mr^2}{M}\,d\varphi.$$

Then we exclude dt from (6) and find:

$$\varphi = \pm\frac{M}{\sqrt{2m}} \int_{r_0}^{r} \frac{dr}{r^2\sqrt{E - U_{\text{eff}}(r)}} + \varphi_0\,. \tag{2.8}$$

It should be noted that the angular velocity $\dot{\varphi} = M/(mr^2)$ has the same sign for the entire orbit, and it is non-zero at the turning points for the radial motion r_i. Therefore,

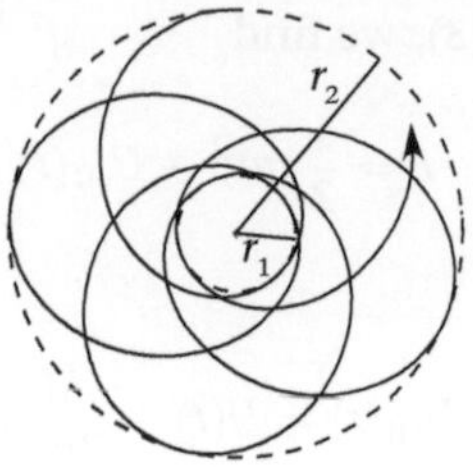

Figure 3 Orbit of a finite motion at $T_\varphi < T_r$

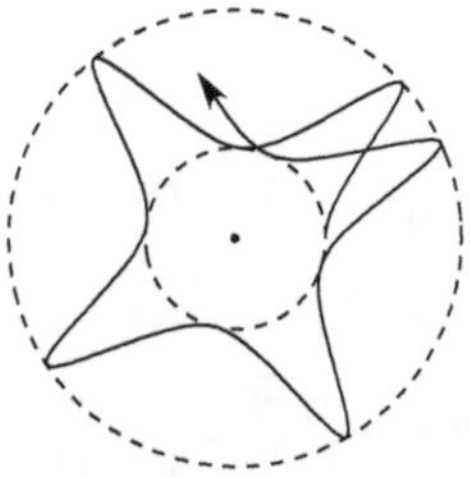

Figure 4 Orbit of a finite motion at $T_\varphi > 4T_r$

the orbits in these points are tangential to a circle of radius r_i. By analogy with the period of the radial motion T_r, we introduce the value T_φ, which is the time during which the particle, starting from the point $r = r_{\min}$, will make a complete revolution, and the angle φ will change by 2π.

The approximate form of the orbits for the finite motion is shown in Fig. 3 for such a function $U(r)$, when during one revolution T_φ a particle performs slightly less than one radial oscillation ($T_\varphi < T_r$) and in Fig. 4 for the case when a particle performs several radial oscillations ($T_\varphi > 4T_r$).

The law of angular momentum conservation (2), (4) can be given a visual geometric meaning using the concept of sectorial velocity. Let the particle move in time dt from the point $\mathbf{r}$ by $d\mathbf{r} = \mathbf{v}\,dt$. Over this time, the radius vector of the particle will turn by the angle $d\varphi$ and will 'sweep out' the elementary sector formed by the radius vector $\mathbf{r}$, the radius vector $\mathbf{r} + d\mathbf{r}$, and the particle's orbit element. Since modulus of the vector production $[\mathbf{r}, \mathbf{r} + d\mathbf{r}] = [\mathbf{r}, d\mathbf{r}]$ is equal to $r^2 d\varphi$ (which is the area of the parallelogram with sides $\mathbf{r}$ and $\mathbf{r} + d\mathbf{r}$), then the length of the vector

$$d\mathbf{S} = \frac{1}{2}\,[\mathbf{r}, d\mathbf{r}]$$

is equal to the area of the sector formed by the vector $\mathbf{r}$, vector $\mathbf{r} + d\mathbf{r}$, and the element of the orbit, i.e. to the area which has been 'swept out' by the radius vector of the particle for the time dt. The vector $d\mathbf{S}/dt$ is called *the sectorial velocity*; it equals

$$\frac{d\mathbf{S}}{dt} = \frac{1}{2}\,[\mathbf{r}, d\mathbf{r}/dt] = \frac{1}{2}\,[\mathbf{r}, \mathbf{v}]. \tag{2.9}$$

The direction of this vector determines the orbit plane, while its length $|d\mathbf{S}/dt| = \frac{1}{2}r^2|\dot{\varphi}|$ determines the area swept out by the radius vector of the particle per unit time.

The sectorial velocity is related to the angular momentum $\mathbf{M}$ by the following relation

$$\mathbf{M} = 2m\,\frac{d\mathbf{S}}{dt}\,. \tag{2.10}$$

Therefore, the conservation of angular momentum means that *the sectorial velocity in the central field is a constant.*

Problem

2.1. A particle falls from a finite distance towards the centre of the field $U(r) = -\alpha/r^n$ with $n \geq 2$. Will it make a finite number of revolutions around the centre? Will it take a finite time to fall towards the centre? Find the equation of the orbit for small r.

§ 3 Kepler's problem

Let us consider the motion of a particle in a potential field

$$U(r) = -\frac{\alpha}{r}\,. \tag{3.1a}$$

Here $\alpha = Gm\,m_S$ for the motion of a planet of mass m in the gravitational field of the Sun (m_S is the mass of the Sun, G – gravitational constant) or $\alpha = e^2$ for the motion of an electron in the electric field of a proton (hydrogen atom).[3]

3.1 *Orbits*

The effective potential energy for a given field

$$U_{\text{eff}}(r) = -\frac{\alpha}{r} + \frac{M^2}{2mr^2} \tag{3.2}$$

is shown in Fig. 5. It is seen from the curve that:

when $E \geqslant 0$, the particle coming from infinity is reflected at the point r_1, consequently going back to infinity;

when $E<0$, the particle experiences radial oscillations in the limited area $r_{\min} \leq r \leq r_{\max}$;

when $E = -m\alpha^2/(2M^2)$, the particle moves along a circle of a radius $r_0 = M^2/(m\alpha)$.

The form of the orbit is determined from Eq. (2.8):

$$\varphi = \pm \int \frac{M}{r^2}\,\frac{dr}{\sqrt{2mE + \dfrac{2m\alpha}{r} - \dfrac{M^2}{r^2}}} + \text{const}. \tag{3.3}$$

If we introduce the dimensionless variable

$$u = \frac{p}{r}\,, \quad \text{where } p = \frac{M^2}{m\alpha}\,, \tag{3.4}$$

[3] In fact, the motion of an electron in the atom is determined not by classical, but by quantum mechanics. The classical description is approximately valid for highly excited (so-called *Rydberg's*) states of the hydrogen atom.

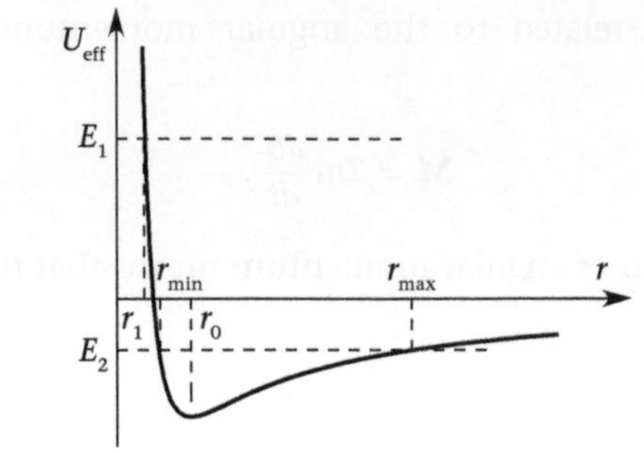

Figure 5 The effective potential energy $U_{\text{eff}}(r) = -\frac{\alpha}{r} + \dfrac{M^2}{2mr^2}$

we obtain

$$\varphi = \mp \int \frac{du}{\sqrt{e^2 - (u-1)^2}} + \varphi_0\,, \quad e = \sqrt{1 + \frac{2EM^2}{m\alpha^2}}\,. \tag{3.5}$$

Further integration is elementary:

$$\varphi = \pm \arccos \frac{u-1}{e} + \varphi_0$$

or

$$r = \frac{p}{1 + e\cos(\varphi - \varphi_0)}\,.$$

Choosing $\varphi_0 = 0$, we have $r = r_{\min}$ for $\varphi = 0$ (for the motion of the planet, this point is called the *perihelion*). As a result, we obtain the equation of the orbit in the form

$$r = \frac{p}{1 + e\cos\varphi}\,, \tag{3.6}$$

where e is the *eccentricity* and p the orbit *parameter*.

Eq. (6) defines the known curves corresponding to the conical cross sections:
a hyperbola for $e > 1$ (when $E > 0$),
a parabola for $e = 1$ (when $E = 0$),
an ellipse for $e < 1$ (when $E < 0$).

When $E = -m\alpha^2/(2M^2)$, the eccentricity is $e = 0$, and the orbit is a circle.
Finally, we point out that the parameter p is equal to the value of r for $\varphi = \pi/2$:

$$p = \frac{M^2}{m\alpha} = r\big|_{\varphi=\pi/2}\,.$$

Note that the parameter of the orbit p is determined by the angular momentum, while the eccentricity is determined by the energy and angular momentum.

It is easy to show that for the repulsive field

$$U(r) = \frac{\alpha}{r}, \tag{3.1b}$$

the equation of the orbit is

$$r = \frac{p}{-1 + e\cos\varphi}, \tag{3.6b}$$

where the quantities e and p are determined by the same formulae (4), (5) as for the field of attraction (1a). In this case, the energy $E > 0$, eccentricity $e > 1$, and the orbit is a hyperbola (our choice of the initial data corresponds to $r = r_{\min} = p/(-1+e)$ for $\varphi = 0$).

3.2 *Elliptical orbit. Kepler's laws*

Let us consider the important case $E < 0$ in more detail. In that case, the orbit is an ellipse with centre C, focus O (where the centre of the gravitational field is located), the major semi-axis $a = CA = (1/2)\,DA$, the minor semi-axis $b = CB$ and the orbit parameter $p = OP$ (Fig. 6). For the motion of the planets, the point A where $r = r_{\min}$ is called the *perihelion*, and the point D where $r = r_{\max}$ is called the *aphelion*. Let us recall the definition of the ellipse. It is the locus of points, the sum of the distances to which from two focuses remains constant (this distance is equal to $2a$). The statement that the planets move along ellipses, in the focus of which the Sun is located, is the essence of *Kepler's first law*.

Since the field under consideration is central, the law of conservation of sectorial velocity is valid for this field. This law can be formulated in the following form: the radius vector of the planet sweeps out equal areas for equal time intervals (*Kepler's second law*).

It is easy to show that the major semi-axis depends only on energy (but not on the angular momentum):

$$a = \frac{1}{2}\,(OA + DO) = \frac{1}{2}\,(r_{\min} + r_{\max}) = \frac{1}{2}\left(\frac{p}{1+e} + \frac{p}{1-e}\right) = \frac{p}{1-e^2} = \frac{\alpha}{2|E|}. \tag{3.7}$$

The distance from the centre of the ellipse C to the focus O is

$$CO = \frac{1}{2}\,(DO - OA) = \frac{1}{2}\,(r_{\max} - r_{\min}) = ae. \tag{3.8}$$

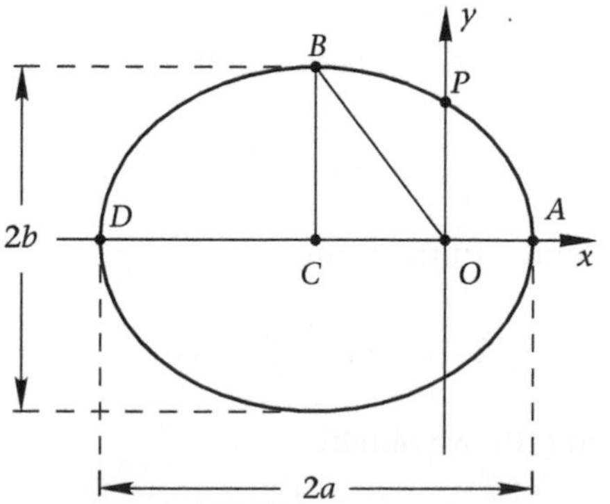

Figure 6 Elements of the elliptical orbit

In a right triangle *COB*, the side *OB* is equal to *a*, so

$$b = CB = a\sqrt{1 - e^2}\,. \tag{3.9}$$

Substituting *e* from (5), we see that *b* depends not only on the energy, but also on the angular momentum:

$$b = \frac{M}{\sqrt{2m|E|}}\,. \tag{3.10}$$

Finally, let us write the useful relations:

$$r_{\min} = (1 - e)\,a, \quad r_{\max} = (1 + e)\,a, \quad p = (1 - e^2)\,a\,. \tag{3.11}$$

The orbit equation can be represented in a new form by shifting the origin of the Cartesian coordinates to the centre of the ellipse $x = x' - ae$:

$$\left(\frac{x'}{a}\right)^2 + \left(\frac{y}{b}\right)^2 = 1\,,$$

or in a parametric form

$$x' = x + ae = a\cos\xi, \quad y = b\sin\xi\,. \tag{3.12}$$

The angular momentum can be expressed in these terms ξ and $\dot\xi$:

$$M = m(x\dot y - y\dot x) = ma(\cos\xi - e)\,b\,\dot\xi\sin\xi - mb\sin\xi\,a\,\dot\xi\sin\xi =$$
$$= mab(1 - e\cos\xi)\,\dot\xi\,.$$

Integrating this relation, we find the dependence of the parameter ξ on time:

$$\xi - e\sin\xi = \frac{M}{mab}\,(t - t_0)\,, \tag{3.13a}$$

where t_0 is the constant of integration. The time $t = t_0$ corresponds to the moment when the particle passes the perihelion. Eqs. (12) and (13*a*) define (in parametric form) the dependence of the Cartesian coordinates of the particles on time.

The total time of one revolution along the ellipse, *T*, corresponds to the change of the parameter ξ from 0 to 2π. Therefore, the period of the revolution equals

$$T = 2\pi\,\frac{mab}{M}\,,$$

while Eq. (13a) can be presented in the form

$$\xi - e\sin\xi = \frac{2\pi}{T}\,(t - t_0)\,. \tag{3.13b}$$

Taking into account (7) and (10), we obtain

$$T = 2\pi\sqrt{\frac{ma^3}{\alpha}}\,. \tag{3.14}$$

From here follows *third Kepler's law*:

$$\frac{T^2}{a^3} = 4\pi^2 \frac{m}{\alpha} = \frac{4\pi^2}{Gm_S}, \tag{3.15}$$

i.e. for all planets the ratio of the square of the revolution period to the cube of the major semi-axis of the ellipse is the same. Note that the period of revolution depends on the major semi-axis of the ellipse, i.e. only on the energy E, but not on the angular momentum M.

3.3 *The Laplace vector as the additional integral of motion in the Kepler problem*

Laplace discovered an additional integral of motion in the Kepler problem besides the energy and angular momentum. To find this integral, it is convenient to consider the changes over time of the unit vector $\mathbf{n} = \mathbf{r}/r$. When the particle moves, this vector rotates with an angular velocity $\dot{\varphi}$ in the orbital plane, and a small change in this vector $d\mathbf{n}$ is perpendicular to the vector itself and to the vector of the angular momentum $\mathbf{M}$. Thus, the rate of this vector change, directed along the vector $[\mathbf{M}, \mathbf{n}]$, coincides with $\dot{\varphi} = M/(mr^2)$ in magnitude, i.e.

$$\frac{d\mathbf{n}}{dt} = \left[\frac{\mathbf{M}}{mr^2}, \mathbf{n}\right]. \tag{3.16}$$

Since the equation of motion in the field $U(r) = -\alpha/r$ has the form

$$m\frac{d\mathbf{v}}{dt} = -\frac{\alpha}{r^2}\mathbf{n},$$

Eq. (16) can be rewritten as

$$\frac{d\mathbf{n}}{dt} = -\left[\frac{\mathbf{M}}{\alpha}, \frac{d\mathbf{v}}{dt}\right]$$

or

$$\frac{d}{dt}\left([\mathbf{v}, \mathbf{M}] - \alpha\frac{\mathbf{r}}{r}\right) = 0.$$

As a result, in the attractive field $U(r) = -\alpha/r$, there exists the additional integral of motion – *the Laplace vector* (sometimes also called the Runge–Lenz vector)

$$\mathbf{A} = [\mathbf{v}, \mathbf{M}] - \alpha\frac{\mathbf{r}}{r}. \tag{3.17a}$$

It is obvious from this proof that in the repulsive field $U(r) = \alpha/r$ the Laplace vector has the form

$$\mathbf{A} = [\mathbf{v}, \mathbf{M}] + \alpha\frac{\mathbf{r}}{r}. \tag{3.17b}$$

To find out the visual interpretation of the vector $\mathbf{A}$, we consider the scalar product of vectors $\mathbf{r}$ and $\mathbf{A}$. By designating through φ the angle between these vectors, we find

$$\mathbf{r}\mathbf{A} = rA\cos\varphi = \mathbf{r}\,[\mathbf{v}, \mathbf{M}] - \alpha r = \frac{M^2}{m} - \alpha r = \alpha(p - r),$$

where $p = M^2/(m\alpha)$ or

$$r = \frac{p}{1 + (A/\alpha)\cos\varphi}. \tag{3.18}$$

Comparing this expression with (6), we immediately establish that the vector **A** is directed from the centre of the field to the point $r = r_{\min}$ (to the perihelion of the planet), and the modulus of this vector is proportional to the eccentricity:

$$|\mathbf{A}| = \alpha e. \tag{3.19}$$

Thus, in the Kepler problem, there appear seven integrals of motion: the energy E, three projections of vector **M** and three of vector **A**. However, only five of them are independent, since the modulus of the vector **A**, according to (19) and (5), is determined by the energy and angular momentum, and the plane in which this vector lies is orthogonal to vector **M**. A possible set of five independent integrals of motion is as follows: E and **M** give four integrals and determine the orbital plane and parameters of the ellipse, but not its orientation on the plane. The fifth independent integral is the direction of the vector **A**, which determines the position of the orbital perihelion.

In conclusion, note that the presence in the Kepler problem of an additional integral of motion — the Laplace vector — leads to additional *hidden* or *dynamical* symmetry (see § 41).

Also note that the arbitrary function of the known integrals of motion in any problem is an integral of motion itself. Therefore, whenever an integral of motion is found, one must check whether or not it is reducible to the known integrals of motion.

Problem

3.1. A spacecraft is moving in a circular orbit of the radius R around the Earth. A body, whose mass is negligible in comparison to the mass of the spacecraft, is thrown from the spacecraft with relative velocity v, directed towards the centre of the Earth. Find the orbit of the body. Consider the Earth as a uniform ball.

This problem is formulated based on a real incident: during a spacewalk, cosmonaut A. Leonov threw the plug from the camera in the direction of the Earth — see [4], § 8.

§ 4 Perihelion precession under the perturbation $\delta U(\boldsymbol{r})$

When a planet moves in the field of the Sun, it experiences not only the Newtonian force corresponding to potential energy $U(r) = -\alpha/r$, but also the smaller forces: from other planets, from large satellites (like the Moon near the Earth) and so on. In addition, it is sometimes necessary to take into account the relativistic corrections. Therefore, the question arises of constructing a perturbation theory for an approximate account of such forces. We consider a simple example when the perturbative field is central. Let the particle move in the field

$$U(r) = -\frac{\alpha}{r} + \delta U(r),$$

where $\delta U(r)$ is such a small perturbative central field. The equation of motion is

$$m\dot{\mathbf{v}} = -\frac{\alpha \mathbf{r}}{r^3} - \delta U'(r)\frac{\mathbf{r}}{r}$$

with

$$\delta U'(r) = \frac{d(\delta U)}{dr}\,.$$

In this problem, the energy and angular momentum are conserved. Per singular period of the radial oscillation, the particle's orbit is almost identical to an unperturbed ellipse. The only difference between that and the ordinary ellipse is that the former slowly rotates in the orbit plane. Therefore, it is convenient to look for the change of the Laplace vector $\mathbf{A}$ averaged over the period of motion, since just this vector can give us the visual information about the particle orbit. In this problem, the vector $\mathbf{M}$, as before, is conserved, but the time derivative of the vector $\mathbf{A}$ is no longer equal to zero:

$$\frac{d\mathbf{A}}{dt} = [\mathbf{v}, \mathbf{M}] - \alpha\frac{d}{dt}\frac{\mathbf{r}}{r} = -[\mathbf{G}, \mathbf{M}]\,, \quad \mathbf{G} = \frac{\delta U'}{m}\frac{\mathbf{r}}{r}\,.$$

Averaging this equation over one period of the particle's motion along the ellipse, we get

$$\left\langle \frac{d\mathbf{A}}{dt} \right\rangle = -[\langle \mathbf{G} \rangle, \mathbf{M}]\,,$$

where we introduce the notation

$$\langle F(t) \rangle = 1\frac{1}{T}\int_0^T F(t)\,dt$$

(here T is the period of motion of the particle along the ellipse). Obviously, the vector $\langle \mathbf{G} \rangle$ directed along the major semi-axis of the ellipse, that is, parallel to the vector $\langle \mathbf{A} \rangle$:

$$\langle \mathbf{G} \rangle = C\,\langle \mathbf{A} \rangle\,,$$

where

$$C = \left\langle \frac{\delta U'}{m}\frac{\mathbf{rA}}{rA^2} \right\rangle = \frac{1}{T}\int_0^T \frac{\delta U'(r(t))}{m}\frac{\mathbf{r}(t)\mathbf{A}}{r(t)A^2}\,dt\,.$$

Using equation

$$\mathbf{rA} = rA\cos\varphi, \quad dt = \frac{mr^2}{M}\,d\varphi, \quad A = \alpha e\,,$$

we obtain

$$C = \frac{1}{\alpha e M T}\int_0^{2\pi} r^2 \delta U'(r)\cos\varphi d\varphi\,, \tag{4.1}$$

where the dependence of r on φ is determined by the unperturbative motion

$$r \equiv r(\varphi) = \frac{p}{1 + e\cos\varphi}\,. \tag{4.2}$$

As a result, we find (omitting the averaging sign)

$$\frac{d\mathbf{A}}{dt} = C[\mathbf{M}, \mathbf{A}],$$

whence it follows that the vector **A** rotates with a small angular velocity

$$\boldsymbol{\omega} = C\,\mathbf{M}. \tag{4.3}$$

From (3) and (2) it follows that the displacement of the perihelion during one period is equal to

$$\delta\varphi = \omega T = \frac{1}{\alpha e}\int_0^{2\pi} r^2\,\delta U'(r)\cos\varphi\,d\varphi, \tag{4.4}$$

where $r \equiv r(\varphi)$ was defined in (2).

In particular, for the perturbation in the form

$$\delta U(r) = \frac{\beta}{r^2} \tag{4.5}$$

we obtain $\delta U'(r) = -2\beta/r^3$ and

$$\delta\varphi = -\frac{2\beta}{\alpha e}\int_0^{2\pi}\frac{1}{r}\cos\varphi\,d\varphi = -\frac{2\beta}{\alpha e}\int_0^{2\pi}\frac{1+e\cos\varphi}{p}\cos\varphi\,d\varphi = -\frac{2\pi\beta}{\alpha p}. \tag{4.6}$$

Figure 7 shows the orbit of motion in the Coulomb field with this perturbation for $\beta<0$. The particle starts from the aphelion point and performs five radial oscillations.

Problem

4.1. The particle moves in the field $U(r) = -\alpha/r$. Find the displacement of the perihelion during one period for a perturbation of the form $\delta U = \gamma/r^3$.

Such a perturbation approximately describes the influence of the Moon on the motion of the Earth in the field of the Sun — see problem 2.25 in [3].

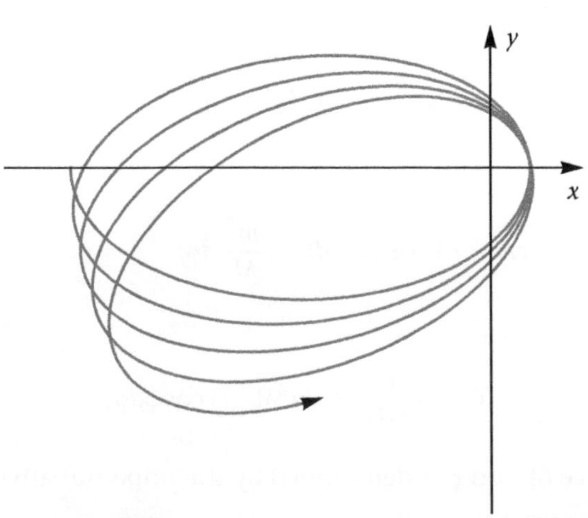

Figure 7 Orbit of particle in the field $U(r) = -\frac{\alpha}{r} + \frac{\beta}{r^2}$ at $\beta<0$

§ 5 Motion in the central field $U(r) = -\frac{\alpha}{r} + \frac{\beta}{r^2}$

Consider the motion of a particle in a field

$$U(r) = -\frac{\alpha}{r} + \frac{\beta}{r^2}.$$

This is the same problem as described in the previous section; however, this time, we will present not an approximate but an exact solution. Such a field (for $\beta<0$) naturally arises when one takes into account the relativistic effects for the motion of a planet in the field of the Sun (see detail in § 48.2). Besides, in this case, it is easy to obtain a solution in the analytical form by a simple reduction that the given problem to the Kepler problem, provided that the quantity

$$\tilde{M}^2 = M^2 + 2m\beta > 0.$$

Indeed, in this case, the effective potential energy for the considered field

$$U_{\rm eff}(r) = -\frac{\alpha}{r} + \frac{\tilde{M}^2}{2mr^2}, \quad \tilde{M} = \sqrt{M^2 + 2m\beta} \tag{5.1}$$

has qualitatively the same form as $U_{\rm eff}(r)$ on Fig. 5. The equation of motion is determined from Eq. (2.8):

$$\varphi = \pm \int \frac{M}{r^2} \frac{dr}{\sqrt{2mE + \dfrac{2m\alpha}{r} - \dfrac{\tilde{M}^2}{r^2}}} + \varphi_0 . \tag{5.2}$$

We rewrite this equation in the form

$$\gamma\varphi = \pm \int \frac{\tilde{M}}{r^2} \frac{dr}{\sqrt{2mE + \dfrac{2m\alpha}{r} - \dfrac{\tilde{M}^2}{r^2}}} + \text{const}, \quad \gamma = \frac{\tilde{M}}{M} = \sqrt{1 + \frac{2m\beta}{M^2}}, \tag{5.3}$$

which differs from Eq. (3.3) for the Coulomb field by replacements $M \to \tilde{M}$, $\varphi \to \gamma\varphi$ only. As a result, we obtain the equation of motion

$$r = \frac{\tilde{p}}{1 + \tilde{e}\cos(\gamma\varphi)}, \tag{5.4}$$

where we introduce quantities

$$\tilde{p} = \frac{\tilde{M}^2}{m\alpha}, \quad \tilde{e} = \sqrt{1 + \frac{2E\tilde{M}^2}{m\alpha^2}}. \tag{5.5}$$

For the case $E<0$ (in this case, $\tilde{e} < 1$), the orbits are shown in Fig. 8. Points A, B, A_1 correspond to the motion from the perihelion A to the aphelion B for the first half-period

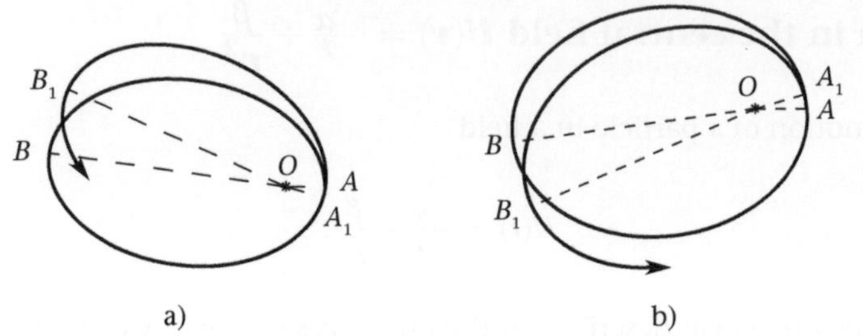

Figure 8 Orbit (5.4) for the case $E<0$: a) when $\beta>0$ and b) when $\beta<0$

of the radial oscillation, and then to the perihelion A_1 for the second half-period of the radial oscillation. The polar angle of a particle changes for one radial oscillation is

$$\Delta\varphi = \frac{2\pi}{\gamma}. \tag{5.6}$$

This means that the time T_φ for one full revolution is

$$T_\varphi = \gamma T_r. \tag{5.6a}$$

The point at which the orbit touches the circle, $r = r_{\min}$, shifted by the angle

$$\delta\varphi = \angle AOA_1 = \frac{2\pi}{\gamma} - 2\pi, \tag{5.7}$$

and $\delta\varphi < 0$ when $\beta>0$ (perihelion shifts clockwise), while $\delta\varphi > 0$ when $\beta<0$ (perihelion shifts counter clockwise).

In the limiting case $2m|\beta|/M^2 = 2|\beta|/(\alpha p) \ll 1$ we obtain

$$\delta\varphi = -\frac{2\pi m\beta}{M^2} = -\frac{2\pi\beta}{\alpha p},$$

which coincides with the result (4.6).

The parameter γ depends on the angular momentum M (see (3)) and can be either a rational or an irrational number. If $\gamma(M)$ is an irrational number, then the orbit is an open curve located in a ring between the circles $r = r_{\min} = \tilde{p}/(1+\tilde{e})$ and $r = r_{\max} = \tilde{p}/(1-\tilde{e})$. This curve fills the ring tightly, passing arbitrarily close to any point on it.

If the angular momentum is such that the parameter $\gamma(M)$ is a rational fraction, $\gamma = n_1/n_2$, and $n_{1,2}$ are integers, then the orbit is a closed curve: the particle returns to the starting point on the orbit after performing n_1 radial oscillations and n_2 full revolutions.

Finally, note that the situation changes qualitatively when $\tilde{M}^2 = M^2 + 2m\beta < 0$; in this case there is a possibility of falling to the centre of the field (for more detail see problem 2.4 in [3]).

The considered example is typical for the motion in the central field, which we have divided into two motions: over the angle φ and over the radius r. In their finite motion, both of them are periodic, but the periods T_φ and T_r are, generally speaking, incommensurable and, therefore, the orbits of a finite motion are typically not closed. In other words, for the finite motion in an arbitrary central field, the particle moves in such a way that its angle of rotation for the period of one radial oscillation $\Delta\varphi$ in the general case (for arbitrary values of the angular momentum and energy, admissible for the finite motion) is incommensurable with the angle of total revolution 2π, i.e. the ratio $2\pi/\Delta\varphi$ is an irrational number, and thus the orbit is not a closed curve. It can be shown (see, for example, [4], § 8) that there are only two exceptions: the Coulomb field $U(r) = -\alpha/r$ in which $2\pi/\Delta\varphi = 1$ (or $T_\varphi = T_r$), and the field of the isotropic oscillator $U(r) = kr^2/2$, in which $2\pi/\Delta\varphi = 2$ (or $T_\varphi = 2\,T_r$) – see below, § 6. In those fields, the orbits of the finite motion are closed curves for arbitrary values of the angular momentum, while the energy $E<0$ for the Coulomb field and $E>0$ for the field of an isotropic oscillator. Besides, in the same fields there are additional (that is, besides the energy and angular momentum) integrals of motion and additional symmetry (see § 41 about dynamical symmetry in the Kepler problem).

§ 6 Isotropic oscillator

Another important example of a central field is the potential field of the isotropic oscillator

$$U(r) = \frac{1}{2}kr^2. \tag{6.1}$$

The motion in such a field occurs in a plane perpendicular to the constant vector of the angular momentum **M**, let us call it the *xy*-plane. The orbit of this motion can be found using the general formulae from § 2. It is more convenient, however, to use the equations of motion in the Cartesian coordinates, in which the equations split up

$$\ddot{x} = -\omega^2 x\,, \;\; \ddot{y} = -\omega^2 y\,, \;\; \omega = \sqrt{k/m}\,, \tag{6.2}$$

and the corresponding solutions are well-known:

$$x(t) = A\cos(\omega t + \alpha)\,, \;\; y(t) = B\cos(\omega t + \beta)\,, \tag{6.3}$$

where A, B, α and β are the constants which are defined by the initial conditions.

This solution corresponds to the motion along an ellipse. To demonstrate this, let us get rid of the time t in Eq. (3). To do this, note that

$$\cos(\omega t + \beta) = \cos\delta\,\cos(\omega t + \alpha) - \sin\delta\,\sin(\omega t + \alpha)\,, \;\; \delta = \beta - \alpha\,,$$

and therefore

$$y = x\frac{B}{A}\cos\delta - B\sin\delta\,\sin(\omega t + \alpha).$$

Further using Eq. (3), we obtain

$$\frac{x^2}{A^2} + \frac{1}{B^2\sin^2\delta}\left(y - \frac{B}{A}x\cos\delta\right)^2 = \cos^2(\omega t + \alpha) + \sin^2(\omega t + \alpha) = 1\,. \tag{6.4}$$

That is the equation of the ellipse, the axes of which do not coincide with the axes of x and y.

After a suitable rotation in the xy-plane and a shift in time, we transform (4) to the sum of the squares

$$\frac{x^2}{a^2} + \frac{y^2}{b^2} = 1 \tag{6.5}$$

and present Eq. (3) in the standard form

$$x(t) = a\cos\omega t\,,\quad y(t) = b\sin\omega t\,, \tag{6.6}$$

which corresponds to the initial data $\mathbf{r}_0 = (a,\,0,\,0)$, $\mathbf{v}_0 = (0,\,b\omega,\,0)$. Now it is clear that the orbit of the motion is the ellipse with semi-axes a and b, the centre of which coincides with the centre of the field. The period of revolution $T = 2\pi/\omega$ equals twice the period of radial oscillations $T_r = \pi/\omega$.

A complete set of the integrals that uniquely determine the orbit includes five of them, as in the Kepler problem. As the independent ones, we can choose the angular momentum $\mathbf{M}$ and two energies E_x and E_y of the independent oscillations along the axes x and y. The specified values E_x, E_y, M are sufficient to determine A, B, δ, because $E_x = kA^2/2$, $E_y = kB^2/2$, $M = mAB\omega\sin\delta$. It is clear, that any function of the integrals of motion is itself an integral of motion. Some of these integrals have obvious meanings, like the total energy $E = E_x + E_y$; the meanings of the others, like

$$N = m\dot{x}\dot{y} + kxy\,, \tag{6.7}$$

are less obvious. Conservation of N can be confirmed by the direct differentiation with respect to time, taking into account the equations of motion (2).

If, in addition to field (1), there is a small perturbation δU, then the orbit can change qualitatively. The following two examples for the case of motion in the xy-plane will clearly illustrate the point. If, for example, $\delta U = \frac{1}{2}k_1x^2$, then

$$x(t) = a\cos\omega_1 t\,,\quad y(t) = b\sin\omega t\,,\quad \omega_1 = \sqrt{\frac{k+k_1}{m}}\,. \tag{6.8}$$

In this case, the orbit ceases to be closed and usually fills up the rectangle $|x| \le a$, $|y| \le b$ (Fig. 9).

Another example is the perturbation in the form of a central field: $\delta U = \delta U(x^2 + y^2)$. In that case, the orbit also ceases to be closed, resulting in a precessing ellipse that fills up the area between the two circumferences (Fig. 10).

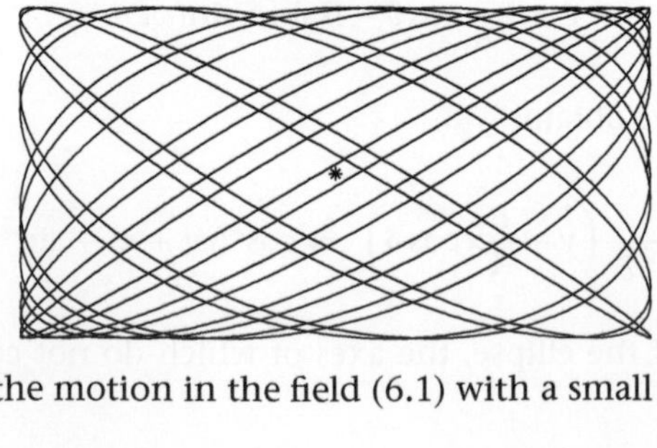

Figure 9 Orbit of the motion in the field (6.1) with a small addition $\delta U = \frac{1}{2}k_1x^2$

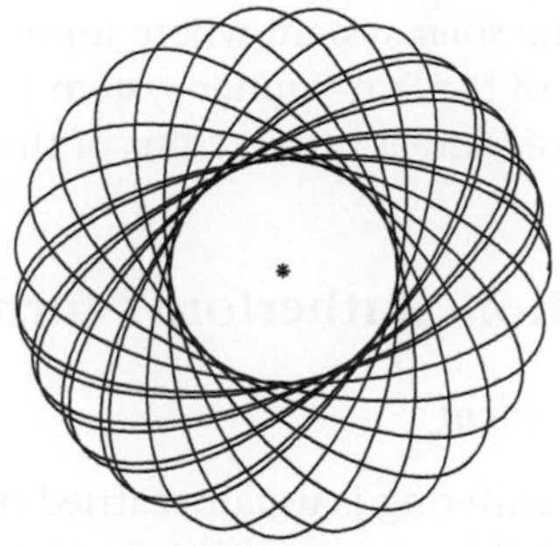

Figure 10 Orbit of motion in the field (6.1) with a small addition $\delta U = \beta/r^4$

Note that in each case the perturbation decreases the number of integrals of motion. In the first case, the angular momentum ceases to be constant due to the disappearance of the central symmetry of the field, while in the second case it is only the total energy, not the separate energies E_x and E_y, that is conserved.

§ 7 The two-body problem

Consider a closed system of bodies consisting of two particles with the potential energy of their interaction in the form $U = U(|\mathbf{r}_1 - \mathbf{r}_2|)$. The equations of motion of this system

$$m_1\ddot{\mathbf{r}}_1 = \mathbf{F}(\mathbf{r}_1 - \mathbf{r}_2) = -\frac{\partial U}{\partial \mathbf{r}_1}, \quad m_2\ddot{\mathbf{r}}_2 = -\mathbf{F}(\mathbf{r}_1 - \mathbf{r}_2) \tag{7.1}$$

can be significantly simplified if we use these new variables: the coordinates of the centre of mass

$$\mathbf{R} = \frac{m_1\mathbf{r}_1 + m_2\mathbf{r}_2}{m_1 + m_2}$$

and the vector of relative distance

$$\mathbf{r} = \mathbf{r}_1 - \mathbf{r}_2$$

instead of the radius-vectors of particles $\mathbf{r}_1$ and $\mathbf{r}_2$. In these variables, equations (1) are separated:

$$\ddot{\mathbf{R}} = 0, \quad m\ddot{\mathbf{r}} = -\frac{\partial U(r)}{\partial \mathbf{r}}, \quad m = \frac{m_1 m_2}{m_1 + m_2}. \tag{7.2}$$

As a result, the two-body problem is reduced to the following two problems: *(i)* the uniform and rectilinear motion of the centre of mass of the system

$$\mathbf{R} = \mathbf{R}_0 + \mathbf{V}t$$

with the initial data $\mathbf{R}(0) = \mathbf{R}_0$, $\dot{\mathbf{R}}(0) = \mathbf{V}$ and *(ii)* to the motion of one particle with *the reduced mass m* under the action of the force

$$\mathbf{F}(\mathbf{r}) = -\frac{\partial U(r)}{\partial \mathbf{r}}.$$

For example, let us consider the solar system where Jupiter's mass is 0.71 of the mass of all planets. The centre of mass of the Sun–Jupiter system is located outside the Sun at the distance $0.07\,R$ above its surface (here R is the radius of the Sun).

§ 8 Scattering cross section. Rutherford's formula

8.1 *Setup of the scattering problem*

The experiment with particle scattering is usually carried out as follows. A beam of particles moving along a certain axis (say, the z-axis), falls on a target, and then the scattered particles are recorded by a detector located at a large distance R from the target. When studying the dependence of the number of scattered particles on the scattering angles and the energy of the incident particles, one can obtain valuable information about the nature of the acting forces, the structure of the target, etc. If the target is sufficiently thin (that is, so that repeated collisions can be neglected) and scattering at the individual scattering centres of the target occurs independently, the problem is reduced to that of the scattering of particles with reduced mass on the potential field $U(\mathbf{r})$. It corresponds to the interaction of a particle from the incident beam with one of the scattering centres of the target. So we come to the following formulation of the scattering problem.

Let the concentration of particles in the incident beam be n, and let their velocity be equal to $\mathbf{v}_\infty = (0, 0, v_\infty)$. Then their density flux is $j = nv_\infty$. After scattering, a certain number of the particles hit the detector in the area $dS = R^2 d\Omega$ with the angular size $d\Omega = \sin\theta d\theta d\varphi$ located at the large distance R from the origin (Fig. 11). The number of the particles $d\dot{N}$ passing through this area in unit time is directly proportional to the value of j. Therefore, the ratio $d\dot{N}/j$ is no longer dependent on the density flux and is determined by the properties of the field of the interaction $U(\mathbf{r})$ and by the initial conditions. If the trajectories of the particles passing through the area dS are traced, we can specify the starting area $d\sigma$ (located perpendicular to the z-axis) through which these particles passed at the initial stage of motion having velocity $\mathbf{v}_\infty$ and impact parameter $\boldsymbol{\rho} = (\rho_x, \rho_y, 0)$. In a real experiment, the impact parameters are usually microscopically small and not directly observed.

The number of particles passing through the area $d\sigma \equiv d^2\rho = d\rho_x\, d\rho_y$ in unit time equals $jd\sigma$ and coincides with the number of the particles $d\dot{N}$ that have passed in unit time through the area dS. Thus, the quantity

$$d\sigma(\theta, \varphi, E) = \frac{d\dot{N}(\theta, \varphi, E)}{j(E)} \tag{8.1}$$

is a convenient characteristic of the scattering process, which may be determined from the experiment by measuring the number of the particles hitting the detector. The value σ

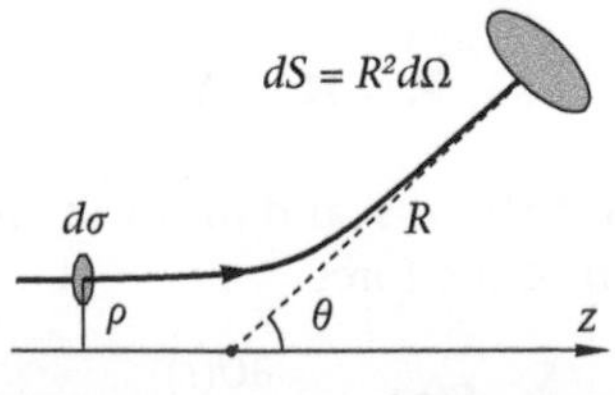

Figure 11 Setup of the scattering problem

obtained after the integration over the scattering angles is called *the total effective scattering cross section* (or *the scattering cross section* for short), while the quantity

$$\frac{d\sigma(\theta, \varphi, E)}{d\Omega} = \left|\frac{d^2\rho(\theta, \varphi, E)}{d\Omega}\right| \tag{8.2}$$

is called *the differential cross section.*[4]

It can be seen from these definitions that both σ and $d\sigma/d\Omega$ are positive values. Additionally, the cross section σ is equal to the total number of the particles scattered by the potential centre in unit time at a unit density flux incident on this centre. If the potential centre is such that the force $\mathbf{F} = -\nabla U(\mathbf{r})$ disappears only at infinity, then all particles from the incident flow will certainly be deflected. It means that the total number of the particles $\dot{N}$ scattered by such a centre is infinite, and, therefore, the total cross-section is also infinite. In particular, if at large distances $U(\mathbf{r}) \sim \alpha/r^n$ with $n>0$, then $\sigma = \infty$.

Let us consider an example of the elastic collision when the particles of the incident beam are balls of radius R_1, and the target particles are also balls of radius R_2. In this case, only the particles with impact parameters $\rho \leq R_1 + R_2$, are scattered, i.e. the total cross section is $\sigma = \pi(R_1 + R_2)^2$.

If the potential field U is central, then the differential cross section is azimuthal angle φ independent and

$$\frac{d\sigma(\theta, E)}{d\Omega} = \left|\frac{d(\pi\rho^2)}{2\pi \sin\theta d\theta}\right| = \frac{\rho(\theta, E)}{\sin\theta}\left|\frac{d\rho(\theta, E)}{d\theta}\right|. \tag{8.3}$$

In this case, in order to calculate the differential cross-section it is sufficient to know $\rho(\theta, E)$, i.e. the impact parameter of the incident particle as a function of its scattering angle and energy.

So far we have considered *the elastic scattering* of particles. Obviously, the concept of a cross section can be extended to the case when particles are falling on the potential centre, or to the case when, upon the particles colliding with the centre, new particles are produced. When describing such processes, *the cross section of falling towards the centre* or *the cross section for inelastic scattering* naturally appear. When we deal with the scattering of particles whose mass is comparable to the mass of the incident particles, we can consider the *the cross section differential in energy transferred to the target particles upon collisions.*

8.2 *Small angle scattering*

Here we compute $\rho(\theta, \varphi, E)$ for the important case of the elastic scattering of fast particles at small angles $\theta \ll 1$. Let $\mathbf{p} = (0, 0, mv_\infty)$ and $\mathbf{p}' = (\mathbf{p}'_\perp, p'_z)$ be the initial and final momenta of a particle, respectively. For the elastic scattering, the magnitudes of these momenta coincide $|\mathbf{p}'| = |\mathbf{p}|$. For small angle scattering, we have

$$\theta \approx \sin\theta = \frac{p'_\perp}{p}, \quad \tan\varphi = \frac{p'_y}{p'_x}, \tag{8.4}$$

where $\mathbf{p}'_\perp = (p'_x, p'_y, 0)$ is transverse to the z-axis component of the vector $\mathbf{p}'$. Thus, to find the scattering angles, it is sufficient to calculate the transverse components of the vector $\mathbf{p}'$.

[4] It is seen from Eq. (2) that the differential as well as the total cross sections have the dimension of area.

In this calculation, we take into account that $\mathbf{p}'_\perp = \Delta\mathbf{p}_\perp$, where $\Delta\mathbf{p} = \mathbf{p}' - \mathbf{p}$ is the total change of the particle's momentum over the entire time of scattering. Since $d\mathbf{p} = \mathbf{F}dt$ due to the Newtonian equation, we get

$$\Delta\mathbf{p} = \int_{-\infty}^{\infty} \mathbf{F}(t)dt = -\int_{-\infty}^{\infty} \frac{\partial U(\mathbf{r}(t))}{\partial \mathbf{r}}\, dt\,.$$

For small-angle scattering, in the right-hand side of this equation, we can substitute the approximate law of motion corresponding to the straight trajectory with an impact parameter $\boldsymbol{\rho}$ and constant velocity $\mathbf{v}_\infty$,

$$\mathbf{r}(t) = \boldsymbol{\rho} + \mathbf{v}_\infty t\,.$$

As a result, we obtain

$$\mathbf{p}'_\perp = -\int_{-\infty}^{\infty} \frac{\partial}{\partial \boldsymbol{\rho}} U(\boldsymbol{\rho} + \mathbf{v}_\infty t)\, dt\,. \tag{8.5}$$

If the potential energy U is the central one,

$$U(r) = U\left(\sqrt{\rho^2 + (v_\infty t)^2}\right),$$

the differential cross section does not depend on the azimuthal angle φ. Using further the replacement $z = v_\infty t$, we get the simple expression for the scattering angle

$$\theta(\rho, E) = \frac{1}{2E}\left|\int_{-\infty}^{\infty} \frac{\partial}{\partial \rho} U\left(\sqrt{\rho^2 + z^2}\right) dz\right|\,. \tag{8.6}$$

From here we find $\rho(\theta, E)$ and $d\sigma/d\Omega$.

Example

We consider the scattering in the field

$$U(r) = \frac{\alpha}{\sqrt{r^2 + a^2}}$$

under the condition $E \gg \alpha/a$. This condition means that the energy of the incident particles E is much higher than the characteristic potential energy α/a. In this case, the scattering occurs only at small angles which are easy to calculate using Eq. (6):

$$\theta = \frac{\alpha\rho}{2E}\int_{-\infty}^{\infty} \frac{dz}{(\rho^2 + z^2 + a^2)^{3/2}} = \frac{\alpha}{E}\frac{\rho}{\rho^2 + a^2}\,.$$

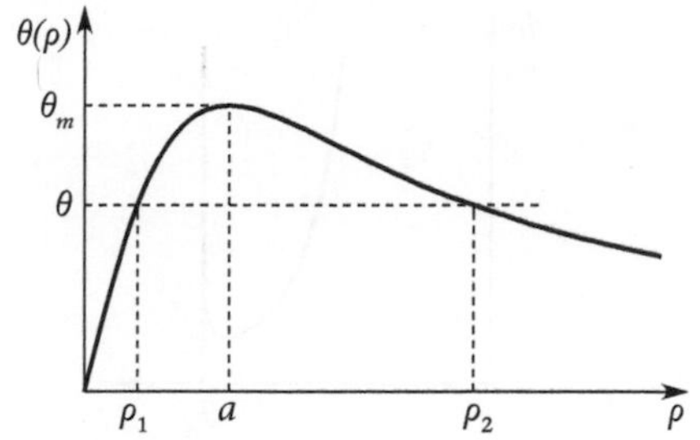

Figure 12 Dependence of $\theta(\rho)$ for small angle scattering in the field $U(r) = \alpha/\sqrt{r^2 + a^2}$

The graph of this function is shown in Fig. 12. The maximum scattering angle θ_m is obtained for $\rho = a$ and, naturally, turns out to be small:

$$\theta_m = \frac{\alpha}{2Ea} \ll 1\,.$$

When calculating the cross section, it is necessary to take into account that the particles flying from two different areas

$$d\sigma_{1,2} = \pi|d\rho_{1,2}^2| = \pm d\rho_{1,2}^2 = \pm\pi\,\frac{d\rho_{1,2}^2}{d\theta}\,d\theta$$

and having two different impact parameters

$$\rho_{1,2} = \left(1 \mp \sqrt{1 - (\theta/\theta_m)^2}\right)\,\frac{\alpha}{2E\theta}$$

hit the same area of the detector. Therefore, we have

$$d\sigma = \pi\left(|d\rho_1^2| + |d\rho_2^2|\right) = \pi d\left(\rho_1^2 - \rho_2^2\right) = \pi d\left[(\rho_1 - \rho_2)(\rho_1 + \rho_2)\right]\,.$$

As a result, we finally obtain

$$\frac{d\sigma}{d\Omega} = \begin{cases} \dfrac{\alpha^2}{E^2\theta^4}\,\dfrac{1 - \theta^2/(2\theta_m^2)}{\sqrt{1 - (\theta/\theta_m)^2}} & \text{when} \quad \theta < \theta_m\,, \\ 0 & \text{when} \quad \theta > \theta_m\,. \end{cases} \tag{8.7}$$

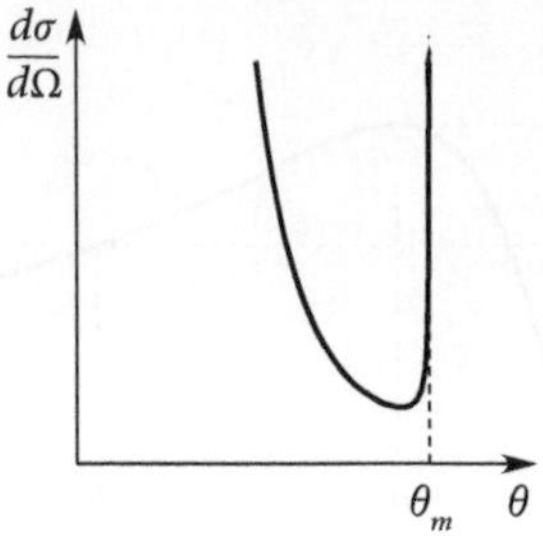

Figure 13 Differential cross section for small angle scattering in the field $U(r) = \alpha/\sqrt{r^2+a^2}$

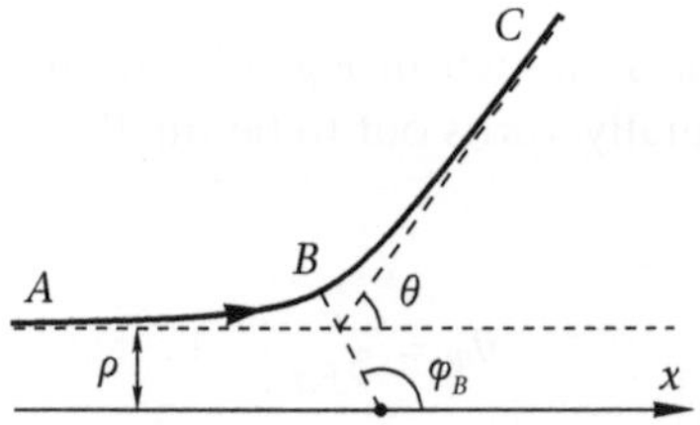

Figure 14 Trajectory of scattering in the Coulomb field $U(r) = \alpha/r$

The dependence of $d\sigma/d\Omega$ on θ is shown in Fig. 13. For $\theta \to 0$, the differential cross section $d\sigma/d\Omega$ increases to infinity:

$$\frac{d\sigma}{d\Omega} \approx \frac{\alpha^2}{E^2\theta^4} \to \infty \text{ when } \theta \to 0. \tag{8.8}$$

The scattering cross section integrated in the range of angles adjoining to $\theta = 0$ is infinite, since the scattering by very small angles corresponds to very large impact parameters.

The differential cross section $d\sigma/d\Omega$ also increases indefinitely for $\theta \to \theta_m$:

$$\frac{d\sigma}{d\Omega} \approx \frac{\alpha^2}{2\sqrt{2}E^2\theta_m^4\sqrt{1-(\theta/\theta_m)}} \to \infty \text{ when } \theta \to \theta_m. \tag{8.9}$$

However, the cross section for scattering of the particles into a range of angles near θ_m is finite, since it corresponds to the finite impact parameters near $\rho = a$. Indeed, the total cross section in the angular interval $\theta_m - \delta < \theta < \theta_m$ is equal to

$$\int_{\theta_m-\delta}^{\theta_m} 2\pi \frac{d\sigma}{d\Omega}\theta d\theta = \frac{\pi\alpha^2\delta^{1/2}}{\sqrt{2}E^2\theta_m^{5/2}}.$$

It is finite and tends to zero as $\delta \to 0$. Such behaviour of the cross section is called *rainbow scattering* (see [5], Ch. 5, § 5). A similar type of cross section behaviour leads to the formation of rainbows in the scattering of light by drops of water. See also problems 3.9, 3.11 from [3] as the examples of rainbow scattering.

8.3 *Rutherford's formula*

We start with the elastic scattering of particles by the Coulomb repulsion field $U(r) = \alpha/r$. The typical trajectory of a particle with energy $E = \frac{1}{2}mv_\infty^2$ and the impact parameter ρ is shown in Fig. 14 as a hyperbola ABC, where points A and C correspond to the initial and final parts of the trajectory, while point B is the minimum distance of the trajectory from the coordinate origin. In the plane of the trajectory (the xy-plane), we introduce polar coordinates r and φ. Now, the equation for the trajectory ABC takes the form (cf. (3.6b))

$$r(\varphi) = \frac{p}{-1 + e\cos(\varphi - \varphi_B)}, \tag{8.10}$$

where the quantities p and e are defined in (3.4), (3.5), and φ_B is a polar angle of point B. The polar angles φ_A and φ_C, corresponding to the points A and C, as well as the scattering angle θ are connected to the angle φ_B by the relations:

$$\varphi_A = \pi, \quad \varphi_C = 2\varphi_B - \pi = \theta .$$

The angle φ_B can be found from the requirement that

$$r(\varphi_A) = \infty \quad \text{or} \quad -1 + e\cos(\pi - \varphi_B) = 0.$$

Taking into account that

$$\varphi_B = \frac{\pi + \theta}{2}, \quad M^2 = (mv_\infty \rho)^2 = 2mE\rho^2, \quad e^2 = 1 + \left(\frac{2E\rho}{\alpha}\right)^2,$$

we find

$$\rho(\theta) = \frac{\alpha}{2E}\cot\frac{\theta}{2} \tag{8.11}$$

and the differential cross section

$$\frac{d\sigma}{d\Omega} = \left(\frac{\alpha}{4E}\right)^2 \frac{1}{\sin^4(\theta/2)}. \tag{8.12}$$

This cross section decreases rapidly with increase of the scattering angle θ (Fig. 15). It is easily demonstrated that Eq. (12) is valid not only for the Coulomb repulsive field $U(r) = \alpha/r$, but also for the Coulomb attractive field $U(r) = -\alpha/r$. At small scattering angles, the result (12) coincides with Eq. (8).

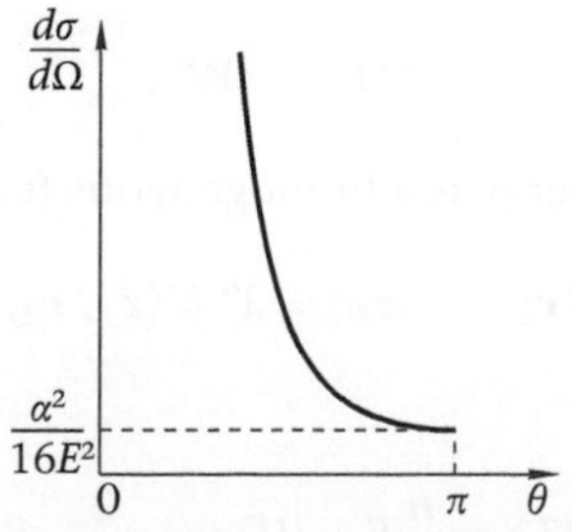

Figure 15 The differential cross section for scattering in the Coulomb field $U(r) = \pm\alpha/r$

Problems

8.1. Find the cross section for the process where a particle having velocity v_∞ at infinity, falls to the surface of the Earth. Take the Earth as a uniform ball of radius R; the acceleration of gravity on its surface is equal to g.

8.2. Find the cross section for the process where a particle falls towards the centre of the field

$$U = \frac{\alpha}{r} - \frac{\beta}{r^2}.$$

How does the answer change when the α sign changes?

8.3. Find the differential cross section for the scattering of particles by an absolutely elastic fixed ball of radius R.

8.4. Find the differential cross section for small angle scattering of fast particles ($E \gg V$) in the field

$$U(r) = \begin{cases} V\left(1 - \dfrac{r^2}{R^2}\right) & \text{when } r < R, \\ 0 & \text{when } r > R. \end{cases}$$

§ 9 Virial theorem

Let us consider a system of N particles whose motion occurs in a limited area and with limited velocities. Let $F(\mathbf{r}_1, \mathbf{r}_2, \ldots \mathbf{r}_N, \mathbf{v}_1, \mathbf{v}_2, \ldots \mathbf{v}_N)$ be some dynamic quantity depending on the coordinates and velocities of the particles. We define the average value of F over a long time interval τ as follows:

$$\langle F \rangle = \frac{1}{\tau} \int_{t_0}^{t_0+\tau} F dt.$$

In the considered system, there are certain connections between the average potential energy $U = U(\mathbf{r}_1, \mathbf{r}_2, \ldots \mathbf{r}_N)$, the average kinetic energy T, and the total energy $E = T + U$ – the so-called *virial theorem.*

Let us call the quantity

$$W = \sum_{a=1}^{N} \frac{\partial U}{\partial \mathbf{r}_a} \mathbf{r}_a \tag{9.1}$$

the virial of the particle system. The virial theorem states that

$$2\langle T \rangle = \langle W \rangle . \tag{9.2}$$

If, in addition, the potential energy is a homogeneous function of degree n, i.e. if

$$U(\lambda\mathbf{r}_1, \lambda\mathbf{r}_2, \ldots \lambda\mathbf{r}_N) = \lambda^n\, U(\mathbf{r}_1, \mathbf{r}_2, \ldots \mathbf{r}_N)\,, \tag{9.3}$$

then

$$\langle T \rangle = \frac{n}{2} E, \;\; \langle U \rangle = \frac{2}{n+2} E. \tag{9.4}$$

To prove this theorem, we start by presenting two auxiliary mathematical statements:

A) If $F(t)$ is a limited function, then

$$\left\langle \frac{dF}{dt} \right\rangle = \frac{1}{\tau} \int\limits_{t_0}^{t_0+\tau} \frac{dF}{dt}\, dt = \frac{F(t_0+\tau) - F(t_0)}{\tau} \to 0 \text{ when } \tau \to \infty\,. \tag{9.5}$$

B) If the potential energy U is a homogeneous function of coordinates and satisfies relation (3), then, according to the Euler theorem on homogeneous functions,

$$\sum_{a=1}^{N} \frac{\partial U}{\partial \mathbf{r}_a} \mathbf{r}_a = n\, U\,. \tag{9.6}$$

To prove this statement, it is sufficient to differentiate equality (3) over λ, and then assume $\lambda = 1$.

Now, let us take the kinetic energy of the particle system

$$T = \frac{1}{2} \sum_{a=1}^{N} m_a \mathbf{v}_a^2$$

and rewrite it in the form

$$2T = \sum_{a=1}^{N} \mathbf{p}_a \mathbf{v}_a. \tag{9.7}$$

Then using the equations of motion,

$$\frac{d\mathbf{p}_a}{dt} = -\frac{\partial U}{\partial \mathbf{r}_a}\,,$$

we transform the item $\mathbf{p}_a \mathbf{v}_a$ in the following way:

$$\mathbf{p}_a \mathbf{v}_a = \frac{d}{dt}(\mathbf{p}_a \mathbf{r}_a) - \frac{d\mathbf{p}_a}{dt} \mathbf{r}_a = \frac{d}{dt}(\mathbf{p}_a \mathbf{r}_a) + \frac{\partial U}{\partial \mathbf{r}_a} \mathbf{r}_a\,.$$

We then substitute this relation into (7). Averaging the resulting expression, we find

$$2\,\langle T \rangle = \left\langle \frac{d}{dt} \sum_{a=1}^{N} \mathbf{p}_a \mathbf{r}_a \right\rangle + \langle W \rangle\,.$$

Taking into account statement (5), we immediately obtain (2).

On the other hand, if the potential energy is the homogeneous function of coordinates, then the Euler theorem (6) allows us to rewrite (2) in the form of the ratio

$$2\langle T \rangle = \langle W \rangle = n\,\langle U \rangle\,.$$

Further, taking into account that $\langle T \rangle + \langle U \rangle = E$, we obtain relations (4).

Some important examples of the application of the virial theorem can be found below.

The field of *the isotropic oscillator* $U(r) = \frac{1}{2}kr^2$ is the homogeneous function with $n = 2$, therefore

$$\langle T \rangle = \langle U \rangle = \frac{1}{2}E. \qquad (9.8)$$

Averaging in this case means the averaging over the period of oscillations.

The Coulomb field $U(r) = -\alpha/r$ is the homogeneous function with $n = -1$; therefore, for the motion along an ellipse (when $E < 0$)

$$\langle T \rangle = -\frac{1}{2}\langle U \rangle = -E. \qquad (9.9)$$

Here averaging means the averaging over the period of revolution.

It is interesting to consider the application of the virial theorem to the *problem of the protostar evolution*. The simplest model of a protostar is a cloud of some monoatomic neutral gas of large mass, which holds together by its own gravitational attraction. For such a star, relations (9) are valid. The kinetic energy of the particles is related to the gas temperature T_g by the well-known formula $\langle T \rangle = \frac{3}{2}NkT_g$, where N is a number of particles in the star and k is the Boltzman constant. The star's energy

$$E = -\langle T \rangle = -\frac{3}{2}NkT_g,$$

and, therefore, its heat capacity $C = dE/dT_g = -\frac{3}{2}Nk$ is *negative*. Note the interesting consequential feature of the star evolution: when radiation is taken into account, the star energy decreases, while its temperature increases over time (see [6], § 21).

Finally, let us consider some cluster of galaxies located very far from any other cosmic objects. The photos of such clusters allow us to conclude that this clusters is in the stationary state, i.e. in the cluster, the single galaxies move, yet their distribution with respect to the centre of the cluster remains unchanged over time. For such a system, Eqs. (9) should be valid. Indeed, these very relations have been found during the initial study of such systems. In the 1930s, however, the more detailed comparison of the kinetic and potential energies of the glowing stars in such clusters showed for the first time the violation of relations (9). That, in turn, led to a hypothesis about the existence of some unknown, invisible, heavy matter, so-called *dark matter*. According to the latest estimates, the mass of this dark matter in the Universe is five times larger than that of the usual visible matter.

Problem

9.1. Find the average kinetic energy of a particle moving in the field $U(r) = V \ln(r/a)$.

CHAPTER II

Lagrangian mechanics

§ 10 Lagrangian equations

10.1 *Lagrangian equations for the non-relativistic particles in a potential field as a covariant notation of Newton's equations*

So far, the motion of a material point (or a particle) has been considered on the basis of Newton's equations. In this chapter we will give another, *Lagrangian form* of these equations, which has several advantages. This section is introductory, so not striving immediately for complete generality of notation and definitions, we will start with a one-dimensional motion. For the motion of a particle of mass m along a straight line x in the potential field $U(x, t)$ the Newton's equation (in some inertial frame of reference) has the form

$$m\ddot{x} = -\frac{\partial U(x,t)}{\partial x}\,. \tag{10.1}$$

Let us introduce a function of three variables x, $\dot{x}$, and t, which is equal to the difference of kinetic and potential energies:

$$L(x,\dot{x},t) = \frac{1}{2}m\dot{x}^2 - U(x,t)\,. \tag{10.2a}$$

This is called the *Lagrangian function or, in short, the Lagrangian.* The equation (1) can be represented as

$$\frac{d}{dt}\frac{\partial L}{\partial \dot{x}} = \frac{\partial L}{\partial x}\,. \tag{10.3}$$

This equation is called the *Lagrangian equation.* In comparison with Newton's equation (1), this one doesn't have any new physical content.

Note one of the advantages of the Lagrangian approach. This form of the equation of motion is convenient, in particular, for the problem of how to pass from the Cartesian coordinates x to any other system of coordinates q, i.e. to the replacement

$$x = x(q,t)\,. \tag{10.4}$$

It turns out that if we make such a change in the Lagrangian

$$L\left(x(q,t), \frac{dx(q,t)}{dt}, t\right) \equiv L'(q,\dot{q},t)\,, \tag{10.5}$$

Lectures on Analytical Mechanics. G. L. Kotkin, V. G. Serbo, A. I. Chernykh, Oxford University Press. © G. L. Kotkin, V. G. Serbo, A. I. Chernykh (2024). DOI: 10.1093/oso/9780198894674.003.0002

then the equation of motion can be represented as

$$\frac{d}{dt}\frac{\partial L'}{\partial \dot{q}} = \frac{\partial L'}{\partial q}, \tag{10.6a}$$

coinciding in form with (3). This type of equations of motion, expressed in terms of the Lagrangian function, is independent of the choice of coordinates. This property is called the equation's *covariance* and can be easily checked by direct calculation (see problem 4.3 from [3]). Another proof will be given in § 11.

10.2 *Generalized coordinates and momenta*

The following statement is easy to prove by direct verification. If we take the Lagrangian for the system of N material points as the difference between the kinetic and potential energies in the Cartesian coordinates in some inertial frame of reference

$$L = T - U(\mathbf{r}_1, \mathbf{r}_2, ..., \mathbf{r}_N, t); \quad T = \frac{1}{2}\sum_{a=1}^{N} m_a \dot{\mathbf{r}}_a^2, \tag{10.2b}$$

then Newton's second law can be written as

$$\frac{d}{dt}\frac{\partial L}{\partial \dot{\mathbf{r}}_a} = \frac{\partial L}{\partial \mathbf{r}_a}, \quad a = 1, 2, \ldots N. \tag{10.3b}$$

Similarly to the one-dimensional case, the Lagrangian (2*b*) with the replacement

$$\mathbf{r}_a = \mathbf{r}_a(q_1, \ldots, q_{3N}, t)$$

can be expressed in terms of $3N$ other coordinates q_i and their derivatives $\dot{q}_i$ (called *generalized coordinates and generalized velocities*):

$$L(q_1, \ldots, q_{3N}, \dot{q}_1, \ldots, \dot{q}_{3N}, t)\,.$$

In this case, the Lagrangian equations have the form (hereinafter, for simplification, the letters q and $\dot{q}$ without an index denote the entire set of generalized coordinates and velocities)

$$\frac{d}{dt}\frac{\partial L(q, \dot{q}, t)}{\partial \dot{q}_i} = \frac{\partial L(q, \dot{q}, t)}{\partial q_i}, \quad i = 1, 2, ..., 3N. \tag{10.6b}$$

In particular, the transitions to curvilinear coordinates, to coordinates in non-inertial frames of reference, to 'collective' coordinates of groups of particles (say, describing the motion of their centre of mass and relative motion), etc., mathematically turn out to be exactly the same and are, therefore, considered according to the standard procedure.

In addition to the generalized velocity $\dot{q}_i$, the *generalized momentum* is introduced, which corresponds to the coordinate q_i and is defined by the relation

$$p_i \equiv \frac{\partial L}{\partial \dot{q}_i}\,. \tag{10.7}$$

If q_i is a Cartesian coordinate, for example, $q_i = x$, then the generalized momentum $p_x = m\dot{x}$ coincides with the x-th component of the usual momentum. In the general case, the generalized coordinate does not necessarily have the dimension of length; respectively, the generalized momentum does not necessarily have the dimension of the product of mass times velocity.

Consider, for example, the motion of a particle in a central field. In that case, it is convenient to choose spherical coordinates r, θ, φ as generalized coordinates.[1] Then we get

$$L = \frac{1}{2} m(\dot{r}^2 + r^2\dot{\theta}^2 + r^2\dot{\varphi}^2 \sin^2\theta) - U(r), \tag{10.8a}$$

$$p_r = m\dot{r}, \quad p_\theta = mr^2\dot{\theta}, \quad p_\varphi = mr^2\dot{\varphi}\sin^2\theta. \tag{10.8b}$$

It is easy to check that the generalized momenta p_r, p_θ, and p_φ are related to momentum $\mathbf{p} = m\mathbf{v}$ and angular momentum $\mathbf{M} = [\mathbf{r}, \mathbf{p}]$ by the equations

$$p_r = (\mathbf{p})_r = \mathbf{p} \cdot \frac{\mathbf{r}}{r}, \quad \mathbf{p}^2 = p_r^2 + \frac{\mathbf{M}^2}{r^2}, \quad p_\varphi = M_z, \quad p_\theta^2 + \frac{p_\varphi^2}{\sin^2\theta} = \mathbf{M}^2. \tag{10.9}$$

Problem

10.1. Write down the components of the particle acceleration vector in a spherical coordinate system.

§ 11 Principle of a least action

11.1 *Hamiltonian principle. Covariance of the Lagrangian equations with respect to replacement of coordinates*

The Lagrange equations are directly related to a particular mathematical problem, namely the problem of the calculus of variations (see Supplement A).

Consider first the one-dimensional case. Let the xy-plane have some class of curves (functions $\tilde{y}(x)$) such[2] that all of them pass through the points $A(x_1, y_1)$ and $B(x_2, y_2)$, i.e. $\tilde{y}(x_1) = y_1$, $\tilde{y}(x_2) = y_2$. Among these functions, we need to find a function $y(x)$, which gives the extreme value to the integral

$$J = \int_{x_1}^{x_2} f(y, y', x)\, dx, \quad y' = \frac{dy}{dx},$$

where $f(y, y', x)$ is a given function of three variables.

[1] If one places the origin of the coordinate system in the centre of the given globe of radius r with the north pole lying on the z-axis, then the polar angle θ is measured along the meridian to the south, and the azimuth angle φ—along the latitude to the east. Denote by $\mathbf{e}_r$, $\mathbf{e}_\theta$ and $\mathbf{e}_\varphi$ mutually orthogonal unit vectors, where $\mathbf{e}_r$ is along the radius vector, $\mathbf{e}_\theta$—along the meridian and $\mathbf{e}_\varphi$—along the latitude. In this case, $\mathbf{r} = \mathbf{e}_r r$, $d\mathbf{r} = \mathbf{e}_r dr + \mathbf{e}_\theta r d\theta + \mathbf{e}_\varphi r \sin\theta d\varphi$, the velocity components $d\mathbf{r}/dt$ are equal to $v_r = \dot{r}$, $v_\theta = r\dot{\theta}$, $v_\varphi = r\dot{\varphi}\sin\theta$, and $\mathbf{v}^2 = \dot{r}^2 + r^2\dot{\theta}^2 + r^2\dot{\varphi}^2\sin^2\theta$.

[2] It is assumed that the functions $\tilde{y}(x)$ are sufficiently smooth.

According to the calculus of variations, the required function $y(x)$ can be found as a solution of the differential equation:

$$\frac{d}{dx}\frac{\partial f}{\partial y'} - \frac{\partial f}{\partial y} = 0\,. \tag{11.1}$$

This equation is called the *Euler equation* of the given variational problem. The quantity

$$\frac{\delta J}{\delta y(x)} \equiv \frac{\partial f}{\partial y} - \frac{d}{dx}\frac{\partial f}{\partial y'}$$

is called the *variational derivative* of J with respect to $y(x)$, and *variation (more precisely, the first variation)* J is the quantity δJ defined by the relation

$$\delta J \equiv \int\limits_{x_1}^{x_2} \frac{\delta J}{\delta y(x)}\,\delta y(x)\,dx\,.$$

Similarly, one can consider the problem of determining the extremum of the integral

$$J = \int\limits_{x_1}^{x_2} f(y_1, ..., y_s;\ y'_1, ..., y'_s; x)\,dx, \tag{11.2}$$

depending on many unknown functions $y_i(x)$ (in that case, it is assumed that these functions are independent). A necessary condition for extremum (1) must be satisfied with respect to each of these functions:

$$\frac{d}{dx}\frac{\partial f}{\partial y'_i} - \frac{\partial f}{\partial y_i} = 0, \quad i = 1, 2, ..., s. \tag{11.3}$$

It is evident that the Euler equations (1), (3) are similar to the Lagrangian equations (10.6a), (10.6b). This makes it possible to formulate the following principle for problems in mechanics, the so-called *principle of a least action (Hamilton's principle)*. Let us formulate it just for arbitrary curvilinear coordinates, although from the discovered similarity of the equations, its validity has been proved so far only in the Cartesian coordinates. Let the system of particles at time t_1 be at point A with coordinates $\left(q_1^{(1)}, q_2^{(1)}, \ldots, q_s^{(1)}\right)$ and at the moment in time t_2 at point B with coordinates $\left(q_1^{(2)}, q_2^{(2)}, \ldots, q_s^{(2)}\right)$. Motion of the particle system between these points occurs according to such a law $q_i(t)$ that the integral

$$S = \int\limits_{t_1}^{t_2} L\left(q_1(t), \ldots, q_s(t), \dot{q}_1(t), \ldots, \dot{q}_s(t), t\right)\,dt \tag{11.4}$$

takes on an extreme value, i.e. so that the variation S turns into zero:

$$\delta S = \sum_{i=1}^{s}\int\limits_{t_1}^{t_2}\left(\frac{\partial L}{\partial q_i} - \frac{d}{dt}\frac{\partial L}{\partial \dot{q}_i}\right)\delta q_i\,dt = 0. \tag{11.5}$$

The quantity S is called the *action*. In doing so, it is assumed that the coordinate variations are independent and satisfy the conditions

$$\delta q_i(t_1) = \delta q_i(t_2) = 0\,, \quad i = 1, 2, \dots, s\,.$$

Due to the independence of δq_i coordinate variations, we obtain equations (10.6b) from Hamilton's principle (5). Thus, the Euler equations of this variational problem are, in fact, the Lagrangian equations of a mechanical system.

The formulated principle allows one to immediately reveal the covariance of the Lagrangian equations under coordinate transformations. Indeed, the transformation of coordinates is reduced to the change of variables in the integral of action (4). The value of the integral itself does not change because of that, and neither do the equations – which determine the extremal value of the integral – change their form.

Hamilton's principle can be taken as the basis of mechanics instead of Newton's equations. The significance of this approach, in particular, is due to the fact that similar variational principles can be formulated in other branches of theoretical physics, such as electrodynamics, quantum mechanics, theory of elementary particles, and so on. A simple example of applying this approach to electromechanical systems is considered in § 22.

The fact that the motion of a particle is given by differential equations (Newton's equations), means that according to the known values of coordinates and particle velocity at some moment t the values of coordinates and velocity are determined at a close moment $t + \delta t$. This situation is familiar to us and seems natural. By the way, this is how one can find the law of particle motion numerically. The variational principle states that the particle moves as if it had tried every possible law of motion and has, in a certain sense, preferred one. Such a 'pursuit of a specific goal' (which will be achieved sometimes) seems not only unusual, but also surprising and not quite understood. Of course, one might think that the coincidence of the equations of motion with the Euler equations of the variational problem is just an accident. And so it would be, if we operate within the framework of classical mechanics. However, in quantum mechanics, as we will see later, the variational principle is related to wave properties of particles (studied in detail in quantum mechanics).

11.2 *Transformation of the Lagrangian function under transformation of coordinates and time*

The equations of motion conserve their form of the Lagrangian equations also in the case when the transformation relates to both coordinates and time. But in this case, the transformation of the Lagrangian function is not reduced to a change of variables, and equality (10.5), generally speaking, is not valid any more.

Consider a transformation $q_i, t \to q_i', t'$ such that

$$q_i = q_i(q_1', \dots, q_s', t'),\ \ t = t(q_1', \dots, q_s', t')\,,\ i = 1, 2, \dots, s\,. \tag{11.6}$$

Here are some examples of time transformation. As t', one can use the 'local time' $t' = t - \lambda x$. The quantity $ct' = \sqrt{(ct)^2 - \mathbf{r}^2}$ is the interval in the theory of relativity; such a choice is allowed to obtain equations of motion in their explicit, relativistically covariant form. From a mathematical point of view, one can forget about the physical sense of the variable t' and interpret transformation (6) as replacement variables in $s+1$-dimensional space.

With this replacement, the integral of action is transformed as follows:

$$S = \int_{t_1}^{t_2} L dt = \int_{t'_1}^{t'_2} L \frac{dt}{dt'} dt' . \tag{11.7}$$

Therefore, it is natural to define the new Lagrangian L' by the following relation:

$$L' = L \frac{dt}{dt'} . \tag{11.8a}$$

Then the action will take the form similar to the original one:[3]

$$S = \int_{t'_1}^{t'_2} L' \left(q', \frac{dq'}{dt'}, t' \right) dt' . \tag{11.9}$$

In this case, the covariance of the Lagrangian equations with respect to transformations (6) is conserved. Namely, if in the old variables the Lagrangian equations had the form (10.6b), then in the new variables and for the new Lagrangian, the same form of equations will be conserved:

$$\frac{d}{dt'} \frac{\partial L'}{\partial (dq'_i/dt')} = \frac{\partial L'}{\partial q'_i}, \quad i = 1, 2, \ldots, s . \tag{11.10}$$

Finally, we note that the Lagrangian equations conserve their form if we multiply the Lagrangian function by a constant factor. If one takes as the Lagrangian not $L = T - U$, but $L' = \lambda L$ (T is kinetic energy, U is potential energy, $\lambda = \text{const}$), we get the same equations of motion, only multiplied by λ. But that is insignificant from the point of view of integrating these equations.

Here we would also like to add the following observation. If two systems move independently of each other, then we can formally combine them into one system consisting of independent parts. It suffices to add their Lagrangian functions. But later there may also appear a need to take into account their interaction. For example, the motion of two planets under the influence of the attraction of the Sun is given by the Lagrangians

$$L_i = \frac{1}{2} m_i v_i^2 - \frac{GMm_i}{r_i}, \quad i = 1, 2,$$

or, if preferred, a single Lagrangian $L = L_1 + L_2$ (here M is mass of the Sun, $m_{1,2}$ are masses of the planets, $\mathbf{r}_{1,2}$ their radius vectors, $\mathbf{v}_{1,2} = \dot{\mathbf{r}}_{1,2}$ their velocities, and G is a constant

[3] In more detail, the relation (10*a*) reads (cf. (10.5))

$$L' \left(q', \frac{dq'}{dt'}, t' \right) = L \left(q(q', t'), \frac{dq}{dt}, t(q', t') \right) \cdot \frac{dt}{dt'} , \tag{11.8b}$$

where

$$\frac{dq_i}{dt} = \frac{dq_i}{dt'} \Big/ \frac{dt}{dt'} , \quad \frac{dq_i(q', t')}{dt'} = \frac{\partial q_i}{\partial t'} + \sum_k \frac{\partial q_i}{\partial q'_k} \frac{dq'_k}{dt'} , \quad \frac{dt(q', t')}{dt'} = \frac{\partial t}{\partial t'} + \sum_k \frac{\partial t}{\partial q'_k} \frac{dq'_k}{dt'} .$$

in the Newton's law of gravitation). We take into account the interaction of planets with each other if we add to L the interaction Lagrangian: $L_{\text{int}} = -Gm_1m_2/|\mathbf{r}_1 - \mathbf{r}_2|$. Obviously, under changes of the form $L \to L' = \lambda L$ it is necessary that the multiplier λ be the same for all terms.

§ 12 Lagrangian function for a particle in an electromagnetic field. Ambiguity in the choice of the Lagrangian function

Let a particle with charge e move in the electromagnetic field, given by the scalar $\varphi(\mathbf{r}, t)$ and vector $\mathbf{A}(\mathbf{r}, t)$ potentials. Electric $\mathbf{E}$ and magnetic $\mathbf{B}$ fields are related to potentials as

$$\mathbf{E}(\mathbf{r}, t) = -\frac{\partial \varphi}{\partial \mathbf{r}} - \frac{1}{c}\frac{\partial \mathbf{A}}{\partial t}, \quad \mathbf{B}(\mathbf{r}, t) = \left[\frac{\partial}{\partial \mathbf{r}}, \mathbf{A}\right], \tag{12.1}$$

where c is the velocity of light. It is easy to show that the Lagrangian equation

$$\frac{d}{dt}\frac{\partial L}{\partial \mathbf{v}} = \frac{\partial L}{\partial \mathbf{r}} \tag{12.2}$$

coincides with the known equation of motion

$$m\dot{\mathbf{v}} = e\mathbf{E} + \frac{e}{c}[\mathbf{v}, \mathbf{B}], \tag{12.3}$$

if we choose the Lagrangian in the form

$$L(\mathbf{r}, \mathbf{v}, t) = \frac{1}{2}m\mathbf{v}^2 - e\varphi + \frac{e}{c}\mathbf{A}\mathbf{v}. \tag{12.4}$$

To do this, it suffices to check that the x-components of Eqs. (2) and (3) coincide. Let's leave that to the reader.

In the Lagrangian function, the terms $\frac{1}{2}m\mathbf{v}^2$ and $e\varphi$ are the usual kinetic and potential energies of the particle, while the last term $(e/c)\mathbf{A}\mathbf{v}$, linear in velocity, is neither kinetic nor potential energy. The generalized momentum is

$$\mathbf{p} = \frac{\partial L}{\partial \mathbf{v}} = m\mathbf{v} + \frac{e}{c}\mathbf{A}. \tag{12.5}$$

It is known that the fields $\mathbf{E}$ and $\mathbf{B}$, and hence the equations of motion of particles in the electromagnetic field, do not change under the gauge transformation of potentials, i.e. when replacing

$$\begin{aligned} \varphi \to \varphi' &= \varphi - \frac{1}{c}\frac{\partial f}{\partial t}, \\ \mathbf{A} \to \mathbf{A}' &= \mathbf{A} + \frac{\partial f}{\partial \mathbf{r}}, \end{aligned} \tag{12.6}$$

where $f = f(\mathbf{r}, t)$ is an arbitrary function of coordinates and time. In the Lagrangian formalism, this leads to the fact that potentials φ, $\mathbf{A}$ and φ', $\mathbf{A}'$, which correspond to the

Lagrangian functions L and L', differ in the total time derivative from the function ef/c:

$$L' = \frac{1}{2}m\mathbf{v}^2 - e\varphi' + \frac{e}{c}\mathbf{A}'\mathbf{v} =$$

$$= \frac{1}{2}m\mathbf{v}^2 - e\varphi + \frac{e}{c}\mathbf{A}\mathbf{v} + \frac{e}{c}\left(\frac{\partial f}{\partial t} + \frac{\partial f}{\partial \mathbf{r}}\mathbf{v}\right) = L + \frac{e}{c}\frac{df(\mathbf{r},t)}{dt}.$$

Therefore, these Lagrangians are physically equivalent.

This example demonstrates that the choice of the Lagrangian function is ambiguous. The following general assertion is true: if one adds the total time derivative of any coordinates and time function $F(q,t)$ to the Lagrangian, then the resulting expression can also be considered as another Lagrangian leading to the same equations of motion. Indeed, if

$$L'(q,\dot{q},t) = L(q,\dot{q},t) + \frac{dF(q,t)}{dt}, \tag{12.7}$$

then the values of the actions S and S' differ only by expressions which do not depend on the choice of trial function,

$$S' = \int_{t_1}^{t_2} L'dt = \int_{t_1}^{t_2} Ldt + \int_{t_1}^{t_2} \frac{dF}{dt}dt = S + F(q^{(2)}, t_2) - F(q^{(1)}, t_1),$$

and therefore the corresponding equations of motion do coincide. From the ambiguity of the Lagrangians follows the ambiguity of generalized momenta:

$$p_i' = \frac{\partial L'}{\partial \dot{q}_i} = \frac{\partial L}{\partial \dot{q}_i} + \frac{\partial}{\partial \dot{q}_i}\frac{dF}{dt} = p_i + \frac{\partial F}{\partial q_i}. \tag{12.8}$$

Discussion about some of the issues related to such ambiguity can be found in [3], problems 4.7–4.9.

Two example of the motion of a charged particle in potential and magnetic fields will be considered later. The case when the potential field is the Coulomb one can be found in the next section, while the case when the potential field corresponds to the linear oscillator or anti-oscillator appears in § 27.

§ 13 Classic Zeeman effect

Here we consider the motion of a particle with charge e in the Coulomb field $U(r) = -\frac{\alpha}{r}$ in the presence of a uniform constant magnetic field $\mathbf{B}$. This is the so-called *Zeeman effect* in the hydrogen atom. What will the motion of a classical particle look like in this case?

13.1 *Charged particle in the Coulomb and magnetic fields*

The Lagrangian of the considered problem reads

$$L(\mathbf{r},\mathbf{v},t) = \frac{1}{2}m\mathbf{v}^2 + \frac{\alpha}{r} + \frac{e}{c}\mathbf{v}\mathbf{A}(\mathbf{r}).$$

We will choose the vector potential in the form

$$\mathbf{A}(\mathbf{r}) = \frac{1}{2}\,[\mathbf{B}, \mathbf{r}].$$

For simplicity, we will consider only motion in the xy-plane perpendicular to magnetic field $\mathbf{B} = (0, 0, B)$. In this case, the Lagrangian in polar coordinates is equal to

$$L(r, \varphi, \dot r, \dot\varphi, t) = \frac{1}{2} m \left(\dot r^2 + r^2\dot\varphi^2\right) - \frac{\alpha}{r} + \frac{eB}{2c} r^2\dot\varphi.$$

This function is time- and φ-independent; therefore, the integrals of motion are the energy E and the generalized momentum p_φ (see § 16):

$$E = \frac{1}{2} m \left(\dot r^2 + r^2\dot\varphi^2\right) - \frac{\alpha}{r}, \quad p_\varphi = mr^2\left(\dot\varphi + \Omega\right), \quad \Omega = \frac{eB}{2mc}.$$

Let us consider the case $p_\varphi > 0$ in more detail. Repeating calculations § 2, we find

$$\dot\varphi = \frac{p_\varphi}{mr^2} - \Omega, \tag{13.1}$$

as well as

$$E = \frac{1}{2} m\dot r^2 + U_{\text{eff}}(r), \quad U_{\text{eff}}(r) = -\frac{\alpha}{r} + \frac{1}{2} mr^2 \left(\frac{p_\varphi}{mr^2} - \Omega\right)^2$$

and the trajectory equation

$$\varphi = \pm\sqrt{\frac{m}{2}} \int \frac{\dfrac{p_\varphi}{mr^2} - \Omega}{\sqrt{E - U_{\text{eff}}(r)}}\, dr. \tag{13.2}$$

Qualitatively, the nature of the motion can be investigated, using plots $U_{\text{eff}}(r)$ (see Fig. 16) and taking into account that $U_{\text{eff}}(r) \to +\infty$ for $r \to 0$ and for $r \to \infty$. From this it follows that the motion is finite for any value of energy, and that the period of radial oscillations is

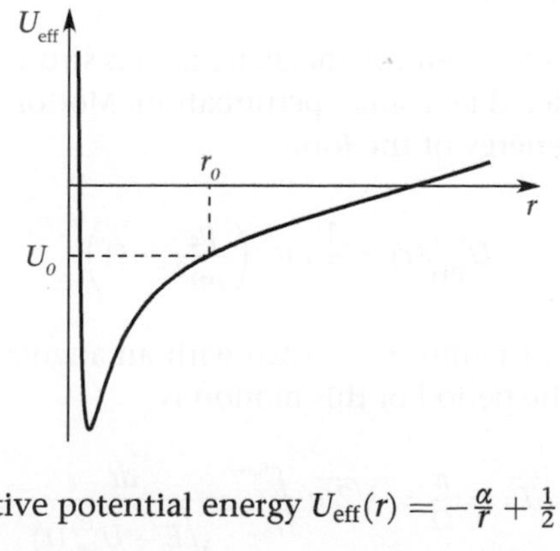

Figure 16 Effective potential energy $U_{\text{eff}}(r) = -\frac{\alpha}{r} + \frac{1}{2} mr^2 \left(\frac{p_\varphi}{mr^2} - \Omega\right)^2$

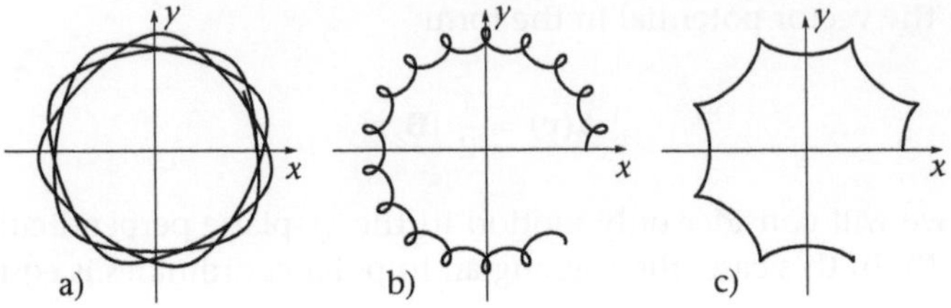

Figure 17 Trajectories described by Eq. (2) for different relations between energy E and U_0: a) $E < U_0$; b) $E > U_0$; c) $E = U_0$

$$T = \sqrt{2m} \int_{r_{\min}}^{r_{\max}} \frac{dr}{\sqrt{E - U_{\text{eff}}(r)}}.$$

At the same time, one should note that, according to (1), the angular velocity $\dot{\varphi}$ changes sign when r passes the value

$$r_0 = \sqrt{\frac{p_\varphi}{m\Omega}}.$$

Thus, in the energy range between the minimum $U_{\text{eff}}(r)$ and the value

$$U_0 \equiv U_{\text{eff}}(r_0) = -\frac{\alpha}{r_0}$$

angular velocity $\dot{\varphi} > 0$, and the trajectory has the form Fig. 17*a*. However, for $E > U_0$, the angular velocity $\dot{\varphi} > 0$ for $r < r_0$ and $\dot{\varphi} < 0$ for $r > r_0$ – see trajectory Fig. 17*b*. Finally, for $E = U_0$ the point $r = r_0$ coincides with the maximum radius, and the trajectory has the form Fig. 17*c*.

Let us consider in more detail two limiting cases of weak and strong magnetic fields. If (i) a magnetic field is weak, and (ii) in the absence of a magnetic field, the particle performs a finite motion in the central field $U(r) = -\alpha/r$, then this problem is easily solved using the Larmor's theorem (see § 20.3).

13.2 *Strong magnetic field. Drift*

More complicated is the case when the magnetic field is strong, while the Coulomb field $U(r) = -\alpha/r$ can be considered as a small perturbation. Motion without $U(r)$ corresponds to the effective potential energy of the form

$$U_{\text{eff}}^{(0)}(r) = \frac{1}{2} mr^2 \left(\frac{p_\varphi}{mr^2} - \Omega\right)^2 \tag{13.3}$$

and occurs along a circle of radius $a = v/(2\Omega)$ with an angular velocity $2\boldsymbol{\Omega} = -e\mathbf{B}/(mc)$ (i.e. clockwise for $e > 0$). The period of this motion is

$$T_0 = \frac{\pi}{\Omega} = \sqrt{2m} \int_{r_{\min}}^{r_{\max}} \frac{dr}{\sqrt{E - U_{\text{eff}}^{(0)}(r)}}. \tag{13.4}$$

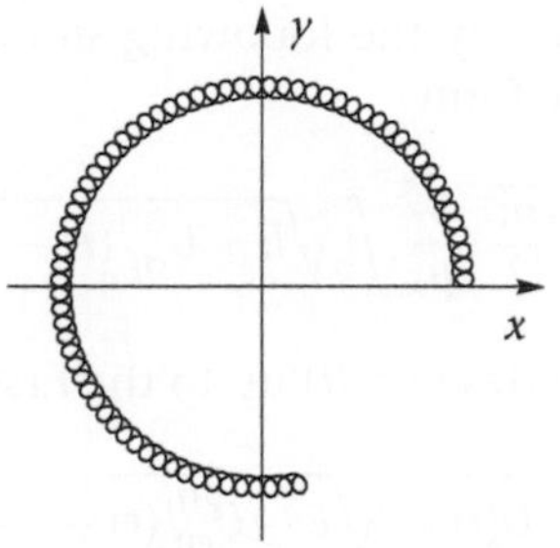

Figure 18 Trajectory of a particle in a strong magnetic field and a weak Coulomb field. It shows the case when a particle in a magnetic field rotates on a circle of small radius a (clockwise), and the centre of this circle slowly drifts along the circumference of a large radius $r_0 = 11\,a$ (counterclockwise)

Accounting for a small field $U(r)$ leads to a systematic displacement of the centre of this circle, called the *drift*. Let us consider in more detail the case when the radius a of the circle is small compared with the distance to the centre of the field $U(r)$. In this case, the motion occurs near the minimum of the effective potential energy (3), i.e. the radius vector of the centre of the circle is equal to r_0 and $r_0 \gg a$. Trajectories of the particle motion for $r_0 = 11a$ are shown in Fig. 18. The drift velocity under these conditions can be found from the following considerations.

Let us consider the case when a charged particles moves in a constant uniform magnetic field **B** in a plane perpendicular to the field. Besides, let there be an additional weak uniform electric field **E** in this plane. It is known (see [10], § 22) that the influence of a weak electric field leads to a slow displacement (drift) of the centre of the circular orbit with a velocity

$$\mathbf{v}_d = c\,\frac{[\mathbf{E}, \mathbf{B}]}{B^2}$$

in a direction perpendicular to **E**. This assertion is easily generalized to the case when, instead of a weak uniform electric field, there is a weak quasi-homogeneous potential field $U(\mathbf{r})$, i.e. such a field that the force $\mathbf{F}(\mathbf{r}) = -\nabla U(\mathbf{r})$ varies only slightly within its circular orbit. In this case, the drift occurs with a velocity

$$\mathbf{v}_d = -c\,\frac{[\mathbf{F}, \mathbf{B}]}{eB^2}\,, \quad \mathbf{F}(\mathbf{r}) = -\nabla U(\mathbf{r})\,.$$

In other words, the centre of the orbit slowly shifts in the direction perpendicular to the force **F**, i.e. along the line $U(\mathbf{r}) = \text{const}$ (the equipotential surface of the field $U(\mathbf{r})$). Such a shift is called *drift*.

In our case, the force is $\mathbf{F} = -\alpha\mathbf{r}/r^3$, and the field equipotential $U(r)$ is a circle of radius r_0. Therefore, the drift is done counterclockwise along a circle of radius r_0 with a velocity

$$v_d = c\,\frac{\alpha}{eBr_0^2} = c\,\frac{2\alpha p_\varphi c}{e^2B^2}. \tag{13.5}$$

If the Coulomb field is a repulsive field, $U(r) = \alpha/r, \;\; \alpha > 0$, then the drift velocity reverses direction.

The result (5) can be confirmed by the following simple calculation. Let us represent the trajectory equation (2) in the form

$$\varphi = \mp\sqrt{\frac{m}{2}}\,\frac{\partial}{\partial p_\varphi}\int\sqrt{E - U^{(0)}_{\mathit{eff}}(r) - U(r)}\,dr$$

and expand the integrand in a series in $U(r)$ up to the first order

$$\sqrt{E - U^{(0)}_{\mathrm{eff}}(r) - U(r)} = \sqrt{E - U^{(0)}_{\mathrm{eff}}(r)} - \frac{U(r)}{2\sqrt{E - U^{(0)}_{\mathrm{eff}}(r)}}\,.$$

The first term on the right-hand side corresponds to a simple rotation along a circle of radius a in a magnetic field in which there is no drift, while the second term leads to a systematic displacement of the centre of the circle (drift) by an angle

$$\delta\varphi = \sqrt{2m}\,\frac{\partial}{\partial p_\varphi}\int_{r_{\min}}^{r_{\max}}\frac{U(r)}{\sqrt{E - U^{(0)}_{\mathrm{eff}}(r)}}\,dr$$

during the time $T_0 = \pi/\Omega$ of one radial oscillation. Since in the case under consideration $r_{\mathrm{min,max}} = r_0 \mp a$ and $a \ll r_0$, the function $U(r)$ can be taken out from under the integral sign at the point $r = r_0$, which gives (taking into account (4))

$$\delta\varphi = \sqrt{2m}\,\frac{\partial}{\partial p_\varphi}U(r_0)\int_{r_{\min}}^{r_{\max}}\frac{dr}{\sqrt{E - U^{(0)}_{\mathrm{eff}}(r)}} = -\frac{\pi}{\Omega}\,\frac{\partial}{\partial p_\varphi}\,\frac{\alpha}{r_0} = \frac{\pi\alpha}{2\Omega^2 r_0}.$$

From here we obtain the angular $\delta\varphi/T_0$ and the linear drift velocity

$$v_d = \frac{\delta\varphi}{T_0}\,r_0 = c\,\frac{\alpha}{eBr_0^2}$$

in accordance with (5).

§ 14 Lagrangian function in the relativistic case

Generalization of the formulae of the previous paragraph to the relativistic case can be done by replacing the non-relativistic Lagrangian (12.4) with the relativistic one:

$$L(\mathbf{r},\mathbf{v},t) = -mc^2\sqrt{1 - \frac{\mathbf{v}^2}{c^2}} - e\varphi + \frac{e}{c}\,\mathbf{A}\mathbf{v}\,. \tag{14.1}$$

The Lagrangian equations (12.2) with such a Lagrangian coincide with the relativistic equations of motion

$$\frac{d}{dt}\frac{m\mathbf{v}}{\sqrt{1 - (v^2/c^2)}} = e\mathbf{E} + \frac{e}{c}[\mathbf{v},\mathbf{B}]\,. \tag{14.2}$$

The generalized momentum is

$$\mathbf{p} = \frac{\partial L}{\partial \mathbf{v}} = \frac{m\mathbf{v}}{\sqrt{1 - (v^2/c^2)}} + \frac{e}{c}\,\mathbf{A}\,. \tag{14.3}$$

In the non-relativistic limit (for $v \ll c$), we obtain from (1):

$$L = -mc^2 + \frac{1}{2}mv^2 - e\varphi + \frac{e}{c}\mathbf{A}\mathbf{v}\,, \tag{14.4}$$

which, up to a constant $-mc^2$, coincides with (12.4).

The Lagrangian (1) corresponds to the action $S = \int_{t_1}^{t_2} Ldt$, which is easy to rewrite in the explicit, relativistically invariant form. In fact,

$$c\sqrt{1-(\mathbf{v}^2/c^2)}dt = \sqrt{(cdt)^2 - (d\mathbf{r})^2} = ds\,, \tag{14.5}$$

where s is the interval and the term

$$(c\varphi - \mathbf{A}\mathbf{v})dt = A_0 dx^0 - \mathbf{A}d\mathbf{r} = A_\mu dx^\mu \tag{14.6}$$

is the scalar product of two 4-vectors, one of which is 4-potential with contravariant $A^\mu = (\varphi, \mathbf{A})$ or covariant $A_\mu = (\varphi, -\mathbf{A})$ components and the other is $dx^\mu = (cdt, d\mathbf{r})$. This action what we get as a result is the Lorentz-invariant quantity

$$S = \int_A^B \left(-mcds - \frac{e}{c}A_\mu dx^\mu\right) \tag{14.7}$$

(here the beginning and end of the integration correspond to the world points $A(ct_1, \mathbf{r}_1)$ and $B(ct_2, \mathbf{r}_2)$, respectively). If the transition $(t, \mathbf{r}) \to (t', \mathbf{r}')$ is the Lorentz transformation, then the invariance of the action implies that $L'dt' = Ldt$. Moreover, the Langrangian $L'(\mathbf{r}', d\mathbf{r}'/dt', t')$ as a function of new variables has exactly the same form (1) as the Langrangian $L(\mathbf{r}, d\mathbf{r}/dt, t)$ as a function of old variables.

Note that the equations of motion for a system of several particles can be obtained using the Lagrangian function containing the interaction energy (10.2b). However, it is impossible to describe a motion of several interacting relativistic particles by generalizing a similar function (1) in a relativistically invariant way. The fact is that under the Lorentz transformations, the time relating to each of the particles will differ for different particles. This can lead to the break of the causality principle in the obtained equations of motion. Let us consider an example when the 'personal' time of particle B, t_B, is less than the time related to particle A, $t_B < t_A$. Then it turns out that the force acting on particle B from the side of particle A at the moment t_B is determined by the position and velocity of particle A at a later moment t_A. It is possible to avoid such impossible situations by including in the theory *fields*, i.e. the electromagnetic field, which provide an interaction of particles.

Side note: obviously, fields in physics also play an independent role. Even the equations, describing the change in the fields over time (such as the Maxwell equations), can also be represented as the Lagrangian equations, albeit somewhat generalized compared to those that are considered in mechanics. These questions, however, are outside of the scope of this course.

Yet, there are cases when it is possible to determine the motion of systems containing both particles and fields by describing them by using just a small number of variables. The electromechanical systems (see § 22 below) are an example of such a system. The Lagrangian function, which approximates, via the electric and magnetic fields, to the interaction of the particles whose velocities are small compared to those of light, can, for example, be found in [11], § 65.

§ 15 The Lagrangian function for systems with ideal holonomic constraints

In classical mechanics, all objects, such as a material point, a rigid body, etc., are idealized, as are the interactions between them. Often such idealized interactions can be described as constraints that restrict the motion of material points and decrease the number of degrees of freedom.

For a large class of systems with the *ideal holonomic constraints*, the definition of which will be given below, the Lagrangian approach turns out to be very effective. Let us start with an example of a mathematical pendulum with a variable length, that is, a particle of mass m suspended in the field of gravity on an inextensible weightless rod, the length of which changes according to the given law $l(t)$ (Fig. 19). We will consider only the motion of the particle in the vertical plane xy, where the x-axis is directed along the gravity force, the y-axis—in the horizontal direction, and the origin of the coordinate systems is at the suspension point. It will be more convenient to specify the particle radius vector $\mathbf{r}$ via the polar coordinates r and φ. The condition of the rod inextensibility means that

$$r = l(t)\,. \tag{15.1}$$

Two forces act on the particle: the force of gravity $m\mathbf{g}$ and the force of the rod tension (or the constraint reaction force) $\mathbf{R}$ directed along $\mathbf{r}$. Note that our system is actually one-dimensional. Indeed, to describe the system, it is sufficient to know $\varphi(t)$ (taking into account condition (1)). The easiest way to find the equations of motion for the coordinate φ is to use the well-known property of the Lagrangian equations, namely, the possibility of writing these equations in any generalized coordinates.

To do this, let us dwell on what is meant by an ideal 'inextensible rod'. Of course, in reality, the rod is simply extremely rigid, so rigid in fact that the magnitude of its deformation is small compared to all other lengths considered in the problem. This longitudinal deformation leads to the appearance of quite a significant force, which, however, in all other respects can be neglected. Let us temporarily suspend such an idealization and introduce the potential energy $\tilde{U}(\tilde{q})$, where

$$\tilde{q} = r - l(t)$$

is an increase in the length of the rod associated with its deformation, and the function $\tilde{U}(\tilde{q})$ very quickly increases with $|\tilde{q}|$ (Fig. 20). Now we can write equations of motion as two-dimensional Lagrangian equations in polar coordinates φ and r. By choosing the variable

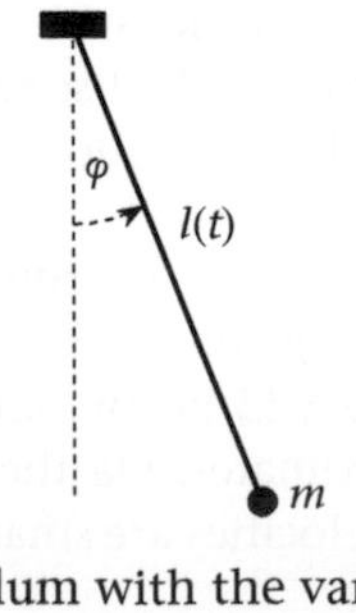

Figure 19 Pendulum with the variable length of rod

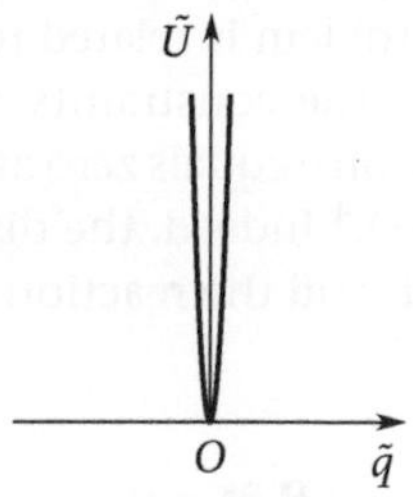

Figure 20 Potential energy of the rod deformation in dependence on the rod length

$\tilde{q}$ instead of r, we can make the further explanation more demonstrable and accessible for the reader.

The Lagrangian function in these variables

$$\tilde{L}(\varphi, \tilde{q}, \dot{\varphi}, \dot{\tilde{q}}, t) = \frac{1}{2} m \left\{ \left[\dot{l}(t) + \dot{\tilde{q}}\right]^2 + [l(t) + \tilde{q}]^2 \dot{\varphi}^2 \right\} + mg\,[l(t) + \tilde{q}] \cos\varphi - \tilde{U}(\tilde{q}) \tag{15.2}$$

leads to the following equations

$$m \frac{d}{dt} \left[(l(t) + \tilde{q})^2 \dot{\varphi} \right] = -mg\,[l(t) + \tilde{q}] \sin\varphi, \tag{15.3}$$

$$m \left[\ddot{l}(t) + \ddot{\tilde{q}} \right] = m\,[l(t) + \tilde{q}]\,\dot{\varphi}^2 + mg\cos\varphi - \frac{d\tilde{U}}{d\tilde{q}}. \tag{15.4}$$

Considering the smallness of $\tilde{q}$, we can put $\tilde{q} = 0$ and $\dot{\tilde{q}} = 0$ in equation (3). As a result, we obtain one equation

$$m \frac{d}{dt} \left[l^2(t) \dot{\varphi} \right] = -mgl(t) \sin\varphi, \tag{15.5}$$

from which one can find dependency $\varphi(t)$ for a given dependence $l(t)$.

Equation (4) formally defines $\tilde{q}(t)$. However, we will not use it, since for the 'inextensible rod' $\tilde{q}(t) = 0$, or, if you like, in accordance with the very definition of 'inextensible rod'.

Finally, the last step would be to put $\tilde{q} = 0$ and $\dot{\tilde{q}} = 0$ at an earlier stage, namely, in the Lagrangian function (2) itself, and discard the term $\tilde{U}$, as unnecessary in the equation (3) for φ. Clearly, such a substitution does not change equation (5) but simplifies the Lagrangian function for a one-dimensional problem:

$$L(\varphi, \dot{\varphi}, t) = \frac{1}{2} m \left[\dot{l}^2(t) + l^2(t) \dot{\varphi}^2 \right] + mgl(t) \cos\varphi. \tag{15.6}$$

Additionally, in this expression, the term $m\dot{l}^2(t)/2$ can be omitted as it does not contain φ and $\dot{\varphi}$.

So, the recipe for obtaining the Lagrangian for a system with constraints has been found: one should introduce (i) the generalized coordinates (considering the constraints!) and (ii) the Lagrangian, that is, the difference between kinetic and potential energies, which should be written under constrained conditions. Moreover, the additional terms of $\tilde{U}$, which characterize very stiff constraints, should be omitted. This course of action allows us to find the equations of the pendulum motion disregarding the reaction forces of constraints.

An alternative approach to this problem is related to the so-called *d'Alembert's principle of virtual displacements*. The force of the constraints reaction $\mathbf{R}$ in our problem has the following property: the work of this force equals zero at any small pendulum displacement $\delta\mathbf{r}$ that does not violate condition (1).[4] Indeed, the displacement $\delta\mathbf{r}$ is directed along the tangent to the circle $r = l(t) = \text{const}$, and the reaction force of the constrain $\mathbf{R}$ is directed along $\mathbf{r}$, so

$$\mathbf{R}\,\delta\mathbf{r} = 0\,. \tag{15.7}$$

In Newton's mechanics, the motion of a particle is determined by the equation

$$m\ddot{\mathbf{r}} = m\mathbf{g} + \mathbf{R}\,, \tag{15.8}$$

which, together with condition (1), allows us to find both the law of motion $\mathbf{r}(t)$ and the force $\mathbf{R}(t)$. It is possible to find the equations of motion for the coordinate φ by substituting $\mathbf{R}$ from Eq. (8) into Eq. (7),

$$(m\ddot{\mathbf{r}} - m\mathbf{g})\,\delta\mathbf{r} = 0\,, \tag{15.9}$$

and applying the constraint condition (1). We leave it to the reader to verify that the above equation of motion for the coordinate φ coincides with Eq. (5).

Now, let us turn from a specific example to the general case. In what follows, we will regard the bodies whose motion we are investigating as 'material points' interacting with each other. For example, under 'absolutely rigid body' we understand the set of material points, the distances between which remain constant. In the similar way, the motion of the system of N material points can be limited by the influence of any rods, surfaces, etc. If all such restrictions are expressed by the conditions

$$F_\alpha(\mathbf{r}_1, ..., \mathbf{r}_N, t) = 0, \quad \alpha = 1, ..., n, \tag{15.10}$$

then we say that *n holonomic constraints* are imposed on the system. The word 'holonomic' indicates that functions (10) do not include velocities. It is assumed that conditions (10) are satisfied due to the fact that, among other things, the constraint forces act on material points. The constraints are called *ideal* if, for any displacements of points that do not violate conditions (10), the total work of all reaction forces is zero. Note that we are not talking about shifts in the process of motion, but about displacements that do not violate condition (1) that is considered at a fixed time.

The pendulum example shows that for a system with ideal holonomic constraints, one can immediately choose generalized coordinates considering the constraints and expressing the Lagrangian function only via those coordinates. In this approach, it is possible to simplify and solve the problem of the system motion, leaving aside the issue of the reaction forces. The benefits of this approach in solving complex problems are hard to overestimate. For more details on holonomic and non-holonomic constraints, see Supplement B.

[4] What is implied is not the displacement during the true motion of the pendulum, but the displacement that does not violate condition (1) for a fixed value of t, i.e. for $l(t) = \text{const}$.

Problems

15.1. Find the Lagrangian function, generalized momentum and energy for the system shown in Fig. 21. Bar mass M can move without friction only along a horizontal straight line, and a particle of mass m can oscillate in a vertical plane on a rod of length l. As the two generalized coordinates, let us choose the Cartesian coordinate X of the bar and the deflection angle φ of a rod from the vertical.

15.2. Two particles of mass M and m are connected by a thread of length l that passes through the hole. The particle of mass M moves in a smooth horizontal plane, while the particle of mass m oscillates vertically in the field of gravity (Fig. 22). Find the Lagrangian function of the system. Consider case when a particle of mass M moves along a trajectory close to a circle (i.e. experiencing small oscillations along the radius). For this case, find the ratio of the frequency of small radial oscillations ω_r to the average angular velocity of motion along the circle $\langle\dot{\varphi}\rangle$ and draw the trajectory under the condition $M = 3m$.

15.3. Find the Lagrangian function and generalized momenta for the system shown in Fig. 23 (double flat pendulum).

15.4. Do the same for the system shown in Fig. 24 (Watt's regulator model). This system rotates in the gravitational field around the vertical axis with a constant angular velocity Ω.

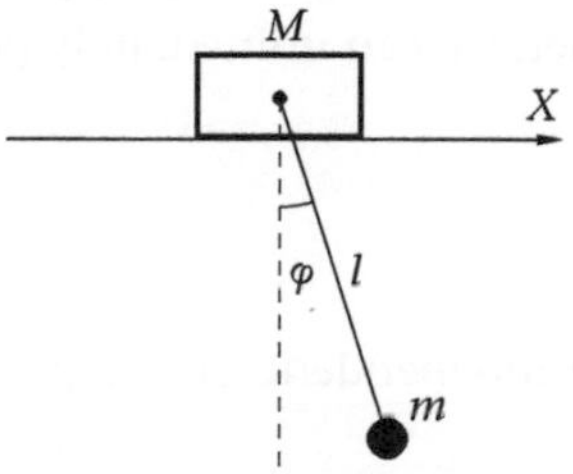

Figure 21 Figure for problem 12.1

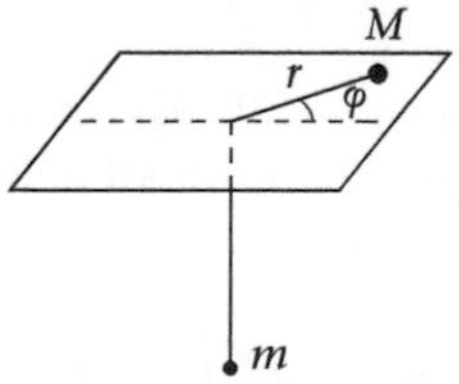

Figure 22 Figure for problem 12.2

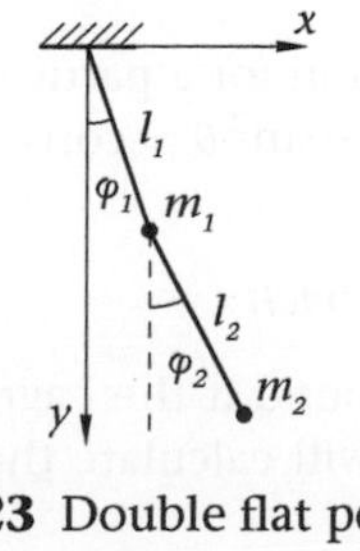

Figure 23 Double flat pendulum

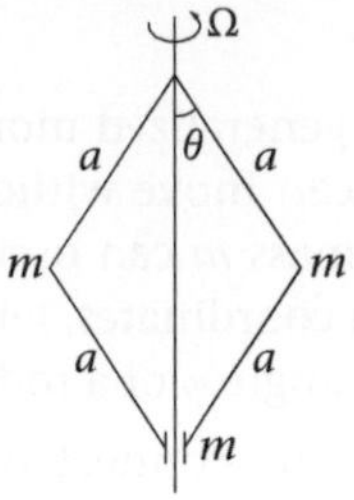

Figure 24 Watt's regulator model

§ 16 Cyclic coordinates. Energy in the Lagrangian approach

Integrals of motion are defined as the functions of generalized coordinates and velocities that remain constant during the motion of a mechanical system. The presence of integrals of motion, as a rule, greatly facilitates the integration of the equations of motion. Thus, the presence of integrals of motion – energies in the one-dimensional case, and energies and angular momentum when a particle moves in a central field – allows us to reduce such problems to quadratures.

The connection between the integrals of motion and the symmetry of the problem will be considered below. Prior to that, let us analyse the following two simple examples: if the Lagrangian function (i) does not depend on some generalized coordinate or (ii) does not *evidently* depend on time, then we can immediately point out the simple integrals of motion.

16.1 *Cyclic coordinates*

Let the Lagrangian function be independent of the generalized coordinate q_k (such a coordinate is called *cyclic*):

$$\frac{\partial L}{\partial q_k} = 0. \tag{16.1}$$

Then the generalized momentum $p_k = \partial L/\partial \dot{q}_k$ corresponding to this cyclic coordinate, is the integral of motion. It immediately follows from the Lagrangian equation

$$\frac{dp_k}{dt} = \frac{d}{dt}\frac{\partial L}{\partial \dot{q}_k} = \frac{\partial L}{\partial q_k} \tag{16.2}$$

and equality (1).

For example, the Lagrangian function for a particle in the central field (10.8) does not depend on φ and, therefore, $p_\varphi = mr^2\dot{\varphi}\sin^2\theta = \text{const}$.

16.2 *Energy in the Lagrangian approach*

Another integral of motion can be found if the Lagrangian function does not evidently depend on time. To prove this, we will calculate the total derivative of the Lagrangian function with respect to time:

$$\frac{dL}{dt} = \sum_i \left(\frac{\partial L}{\partial q_i} \dot{q}_i + \frac{\partial L}{\partial \dot{q}_i} \frac{d\dot{q}_i}{dt} \right) + \frac{\partial L}{\partial t}$$

and rewrite it, taking into account (2), in the form

$$\frac{dL}{dt} = \sum_i \left(\frac{dp_i}{dt} \dot{q}_i + p_i \frac{d\dot{q}_i}{dt} \right) + \frac{\partial L}{\partial t} = \frac{d}{dt} \left(\sum_i p_i \dot{q}_i \right) + \frac{\partial L}{\partial t} . \tag{16.3}$$

Then we introduce the quantity

$$E(t) = \sum_i p_i \dot{q}_i - L = \sum_i \frac{\partial L}{\partial \dot{q}_i} \dot{q}_i - L , \tag{16.4a}$$

called the *energy*. Note the following: the quantities $q_i(t)$ and $\dot{q}_i(t)$ in the right side of Eq. (4*a*) should correspond to the real motion of the system, i.e. E is a function of time. As a result, Eq. (3) can be rewritten in the form

$$\frac{dE(t)}{dt} = -\frac{\partial L}{\partial t} . \tag{16.5a}$$

As a consequence, if the Lagrangian function does not depend evidently on time,

$$\frac{\partial L}{\partial t} = 0, \tag{16.5b}$$

then

$$E(t) = \text{const.} \tag{16.5c}$$

In the non-relativistic case, the Lagrangian function usually contains terms quadratic L_2, linear L_1, and independent L_0 in the generalized velocities:

$$L = L_2 + L_1 + L_0 , \tag{16.6}$$

$$L_2 = \frac{1}{2} \sum_{ik} a_{ik}(q) \dot{q}_i \dot{q}_k; \quad L_1 = \sum_i b_i(q) \dot{q}_i; \quad L_0 = L_0(q, t) .$$

A direct calculation will verify that for such a Lagrangian function the energy is

$$E = L_2 - L_0 . \tag{16.7}$$

In other words, the terms which are quadratic in velocities transfer from the Lagrangian to energy without change, whereas velocity-independent terms change sign, and those terms which are linear in velocities have to be omitted.

As a particular example, let us consider the Lagrangian of a particle in an electromagnetic field (12.4). It contains terms

$$L_2 = \frac{1}{2} m\mathbf{v}^2, \; L_1 = \frac{e}{c}\mathbf{A}\mathbf{v}, \; L_0 = -e\varphi,$$

therefore, the energy reads

$$E = \frac{1}{2} m\mathbf{v}^2 + e\varphi, \tag{16.8}$$

i.e. it is equal to the sum of kinetic and potential energies.

16.3 *Is the energy in the Lagrangian approach equal to the sum of kinetic and potential energies?*

The energy E defined by the formal relation (4*a*) does not always coincide with the sum $T+U$ calculated in the inertial reference frame. Here is a simple example of such a situation.

Let us consider a mathematical pendulum, which is a particle of mass m suspended in a gravitational field on a rigid weightless rod of length l. The rod rotates around a vertical axis with a constant angular velocity Ω, while the pendulum oscillates in a vertical plane that rotates with the rod (Fig. 25). The particle is under the influence of the potential force of gravity $m\mathbf{g}$ corresponding to the potential energy

$$U = -mgl\cos\varphi,$$

where φ is the angle of deviation of the pendulum from the vertical in the rotating plane. The kinetic energy of the particle is

$$T = \frac{1}{2} ml^2(\dot\varphi^2 + \Omega^2 \sin^2\varphi).$$

In this example it is easy to understand why the sum

$$T + U = \frac{1}{2} ml^2(\dot\varphi^2 + \Omega^2 \sin^2\varphi) - mgl\cos\varphi \tag{16.9}$$

is not conserved. The particle is under the influence not only of the potential force of gravity, but also of a non-potential reaction force $\mathbf{R}$ from the rod. The component of this

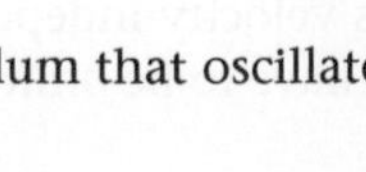

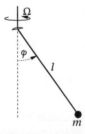

Figure 25 A pendulum that oscillates in a rotating plane

force along the rod $\mathbf{R}_{\parallel}$ is orthogonal to the velocity of the particle and does no work. However, the $\mathbf{R}_{\perp}$ component in the direction transverse to the rod, enforcing the rod to rotate with a constant angular velocity, is the very non-potential force which is responsible for not conserving the sum $T+U$.

In the Lagrangian approach, on the other hand, the above example appears as a system with one degree of freedom and with ideal holonomic constraints. These constraints determine the distance of a particle from the point of suspension of the rod and the angle of the horizontal rotation. Thus, the generalized coordinate φ is the only one needed to completely determine the position of a particle. At fixed time t, superimposed constraints allow the particle to move only in a fixed vertical plane. But in this case, the work of the reaction forces $\mathbf{R}_{\parallel}$ equals zero. This means that the constraints are indeed holonomic and ideal, and the Lagrangian function turns out to be equal to

$$L(\varphi, \dot{\varphi}) = T - U = \frac{1}{2}\, ml^2(\dot{\varphi}^2 + \Omega^2 \sin^2 \varphi) + mgl \cos \varphi\,.$$

This function is time-independent, so energy is conserved and, according to (7), equal to

$$E = \frac{\partial L}{\partial \dot{\varphi}}\dot{\varphi} - L = \frac{1}{2}\, ml^2(\dot{\varphi}^2 - \Omega^2 \sin^2 \varphi) - mgl \cos \varphi\,. \tag{16.10}$$

Note that the term $\frac{1}{2}ml^2\Omega^2 \sin^2 \varphi$, corresponding to the kinetic energy associated with rotation, appears in the energy expression with the ‘minus’ sign, that is, energy (10) is not equal to the sum (9) of the kinetic and potential energies $T+U$ in the inertial reference frame. In § 20.2, we will show that E is the energy in a rotating coordinate system.

16.4 *Ambiguity in the definition of energy*

This section deals with the ambiguity of energy which is associated with the ambiguity in the choice of the Lagrangian function. Let us demonstrate this ambiguity of choice by selecting two Lagrangian functions L and L' which differ in the total time derivative of arbitrary function $F(q, t)$:

$$L'(q, \dot{q}, t) = L(q, \dot{q}, t) + \frac{dF(q, t)}{dt}\,.$$

As indicated in § 12, the corresponding equations of motion coincide. However, the Lagrangian energies E defined by the formula (4*a*), and

$$E'(t) = \sum_i \frac{\partial L'}{\partial \dot{q}_i}\dot{q}_i - L'\,, \tag{16.4b}$$

are different:

$$E'(t) = E(t) - \frac{\partial F(q, t)}{\partial t}\,.$$

For example, if the energy E is conserved, the E' may be time dependent.

For an additional discussion of the issues covered here, see § 17, 18 and [3], and problems 4.15a, 4.18, 4.19.

§ 17 Symmetry and integrals of motion. Noether's theorem

The results of the previous paragraph are the particular manifestation of the general properties of mechanical (and not only mechanical!) systems, namely, *the existence of the integral of motion is a consequence of a certain symmetry of the system.* We are going to use the two earlier mentioned examples to demonstrate this in more detail.

17.1 *Examples*

First, let us consider the motion of a particle in a central field $U(r)$. In this case, the angular momentum of a particle $\mathbf{M}$ is conserved. The existence of this integral of motion is a consequence of the spherical symmetry of the system under consideration. Indeed, in the central field the Lagrangian function (10.8a) does not change under rotation around the z-axis, i.e. under transformation $\varphi \to \varphi + \varepsilon$, where ε is an arbitrary angle of rotation. The consequence of this is the conservation of the generalized momentum $p_\varphi = M_z$. Moreover, in a central field, the direction of the z-axis can be chosen arbitrarily, which means that the vector $\mathbf{M}$ itself is conserved.

The second example is the motion of a particle in a constant potential field $U(\mathbf{r})$. In this case, the energy $E = \frac{1}{2} m\dot{\mathbf{r}}^2 + U(\mathbf{r})$ is conserved. The existence of this integral of motion is connected to the evident time-independence of the Lagrangian function $L(\mathbf{r}, \dot{\mathbf{r}}, t) = \frac{1}{2} m\dot{\mathbf{r}}^2 - U(\mathbf{r})$. That means the Lagrangian does not change under the transformation $t \to t + \varepsilon$, where ε is an arbitrary shift in time.

Now let us consider a somewhat more complicated example: the motion of a charged particle in the field of an infinite uniformly charged helix with the pitch h. If we choose the z-axis along the axis of the helix, this system has a spiral symmetry, i.e. its Lagrangian function does not change under transformations

$$\varphi \to \varphi + \varepsilon, \quad z \to z + \frac{h}{2\pi}\varepsilon,$$

where ε is an arbitrary angle of rotation around the z-axis. If the parameter ε is small, then it follows from this symmetry that

$$\delta L = \frac{\partial L}{\partial \varphi}\varepsilon + \frac{\partial L}{\partial z}\frac{h}{2\pi}\varepsilon = 0. \tag{17.1}$$

Using the Lagrangian equations in the form (16.2),

$$\frac{\partial L}{\partial \varphi} = \frac{d}{dt}\frac{\partial L}{\partial \dot{\varphi}} = \frac{dp_\varphi}{dt}, \quad \frac{\partial L}{\partial z} = \frac{d}{dt}\frac{\partial L}{\partial \dot{z}} = \frac{dp_z}{dt},$$

we get from (1):

$$\frac{d}{dt}\left(p_\varphi + \frac{h}{2\pi}p_z\right) = 0,$$

i.e. in the field under consideration there is an additional integral of motion (note that $p_\varphi = M_z$)

$$M_z + \frac{h}{2\pi}p_z = \text{const}. \tag{17.2}$$

In this example, M_z and p_z are not saved separately, but their combination (2) is conserved.

17.2 *Generalization*

Now from the examples above, it is easy to generalize that if the Lagrangian function does not change under some joint shift in both time and coordinates, then some specific combination of energy and generalized momenta is conserved, although individually neither the generalized momentum nor the energy may be conserved. Namely, let the infinitesimal transformations of time and coordinates have the form

$$t \to t + \varepsilon c_t\,, \quad q_i \to q_i + \varepsilon c_i\,, \quad i = 1, 2, \ldots, s\,, \tag{17.3}$$

where ε is an infinitesimal parameter, and c_t and c_i are some constant values, and let the Lagrangian function of the system stay unchanged under such transformation (up to terms of the order of ε inclusive):

$$\delta L = \varepsilon \left(\frac{\partial L}{\partial t} c_t + \sum_{i=1}^{s} \frac{\partial L}{\partial q_i} c_i \right) = 0\,. \tag{17.4}$$

Then the quantity

$$E c_t - \sum_{i=1}^{s} p_i c_i = \text{const}\,, \tag{17.5}$$

i.e. it is the integral of motion. To prove it, we substitute relations (16.2), (16.5) into Eq. (4) and immediately obtain

$$\frac{d}{dt}\left(-E c_t + \sum_{i=1}^{s} p_i c_i \right) = 0\,,$$

whence the conservation of quantity (5) follows.

17.3 *Noether's theorem*

So far, we have limited ourselves to the invariance of the Lagrangian function with respect to transformation (3), in which c_t and c_i are some constants. It turns out that we can get a more general statement, when in transformation (3) instead of constants c_t and c_i there will be the arbitrary functions of coordinates and time. However, in this case the requirement of invariance is imposed not on the Lagrangian function itself, but on the action. Such a generalization is known as the *Emma Noether's theorem*, detailed below.

Let an infinitesimal transformation of time and coordinates have the form

$$t \to t' = t + \varepsilon h(q, t)\,, \quad q_i \to q_i' = q_i + \varepsilon f_i(q, t)\,, \quad i = 1, 2, \ldots, s\,, \tag{17.6}$$

where ε is an infinitesimal parameter, and, under such a transformation, let the form of the action stay unchanged (up to terms of the order of ε inclusive)[5]

$$\int_{t_1}^{t_2} L\left(q, \frac{dq}{dt}, t\right) dt = \int_{t_1'}^{t_2'} L\left(q', \frac{dq'}{dt'}, t'\right) dt'\,. \tag{17.7}$$

[5] Note that in the left and the right sides of equality (7) are *the same function L*, but of different arguments.

Then the quantity

$$Eh - \sum_{i=1}^{s} p_i f_i = \text{const}\,, \tag{17.8}$$

i.e. it is the integral of motion. A simple proof of this theorem will be given in § 46.3.

It is possible to assign another form to these relations, which sometimes is more convenient in applications. Here is how it can be done. First, let use new notations

$$\varepsilon h(q,t) = \delta t, \quad \varepsilon f_i(q,t) = \delta q_i.$$

Then, if under the transformation

$$t \to t + \delta t, \quad q_i \to q_i + \delta q_i\,, \ \ i = 1, 2, \ldots, s \tag{17.9}$$

the action does not change its form (up to terms of the order of δt and δq_i inclusive), the quantity

$$E\delta t - \sum_{i=1}^{s} p_i \delta q_i = \text{const}\,. \tag{17.10}$$

Noether's theorem is, in essence, a singular proof of the various conservation laws which follows from a certain symmetry of the system. The fact that a similar theorem also holds in field theory renders Noether's theorem increasingly important (see [7, 8]).

To illustrate the application of Noether's theorem, let us consider the motion of a particle in a dipole field. In this case, the Lagrangian function is equal to

$$L(\mathbf{r}, \mathbf{v}) = \frac{1}{2} m\mathbf{v}^2 - U(\mathbf{r}), \quad U(\mathbf{r}) = \frac{\mathbf{ar}}{r^3}\,, \tag{17.11}$$

where the constant vector **a** equals the product of the electric dipole moment and the particle charge. This Lagrangian function is evidently time-independent and stays unchanged under rotation around vector **a**. Therefore, both the energy $E = \frac{1}{2}m\mathbf{v}^2 + U(\mathbf{r})$ and the momentum's projection on the dipole direction $m[\mathbf{r}, \mathbf{v}]\mathbf{a}$ are conserved. However, this system also has an additional symmetry.

It is easy to see that relation (7) is valid for the Lagrangian (11) under the *similarity transformation*

$$\mathbf{r} \to \mathbf{r}' = \lambda \mathbf{r}, \quad t \to t' = \lambda^2 t\,, \tag{17.12}$$

where λ is an arbitrary parameter, since

$$L\left(\mathbf{r}', \frac{d\mathbf{r}'}{dt'}\right) = L\left(\lambda \mathbf{r}, \frac{\mathbf{v}}{\lambda}\right) = L(\mathbf{r}, \mathbf{v}) \frac{1}{\lambda^2} = L(\mathbf{r}, \mathbf{v}) \frac{dt}{dt'}\,.$$

Noether's theorem allows us to find another integral of motion related to the symmetry under the similarity transformation. As the parameter λ we take $\lambda = 1 + \varepsilon$, then

$$\delta \mathbf{r} = \varepsilon \mathbf{r}, \quad \delta t = 2\varepsilon t$$

and from (10) it follows that

$$2Et - m\mathbf{vr} = \text{const} \equiv C_1 . \tag{17.13}$$

Using this integral of motion, it is easy to find the dependence $r(t)$. Indeed, using

$$C_1 = 2Et - m\dot{r}r = 2Et - \frac{m}{2}\frac{dr^2}{dt},$$

we get

$$r(t) = \sqrt{\frac{2}{m}(Et^2 - C_1 t) + C_2},$$

where $C_{1,2}$ are constants determined by the initial conditions.

When obtaining the integral of motion (13), what is essential is the very fact that $U(\lambda\mathbf{r}) = U(\mathbf{r})/\lambda^2$ and not the specific type of potential energy (11) per se. Therefore, the same integral of motion (13) also holds for a particle moving in the central field $U(r) = \beta/r^2$, as well as when a particle moves in the field of the magnetic monopole (see [3], problem 4.23), and so on.

Admittedly, the formulation of Noether's theorem given here is a variant adapted for problems of classical mechanics, whereas the original Noether's theorem was designed for continuous systems. It is also worth mentioning that mechanical systems, possessing this or that symmetry, represent an exception rather than the rule. Moreover, discovering a definite symmetry is often quite a complicated and creative problem. On the contrary, once the symmetry of a system is established, the search for the integrals of motion using Noether's theorem appears as a simple standard procedure. Some of such interesting examples are considered in [1], § 48.

In conclusion, please remember that when the total derivative with respect to time of the function of coordinates and time is added to the Lagrangian function, the Lagrangian equations do not change. Consequently, a somewhat more general theorem is valid: if under transformation (6) the form of the Lagrangian function varies by no more than the total derivative of the function of coordinates and time

$$\int_{t_1}^{t_2} L\left(q, \frac{dq}{dt}, t\right) dt = \int_{t_1'}^{t_2'} \left[L\left(q', \frac{dq'}{dt'}, t'\right) + \varepsilon\frac{dF(q', t')}{dt'}\right] dt', \tag{17.7a}$$

then the integral of motion is equal to

$$Eh - \sum_{i=1}^{s} p_i f_i - F = \text{const}. \tag{17.8a}$$

Problems

17.1. Find integrals of motion for a particle moving in the traveling wave field $U(\mathbf{r}, t) = U(\mathbf{r} - \mathbf{V}t)$, where $\mathbf{V}$ is a constant vector.

17.2. Find integrals of motion for a particle in a homogeneous constant magnetic field $\mathbf{B}$ if the vector potential is given as:

a) $A_x = A_z = 0$, $A_y = xB$,

b) $\mathbf{A} = \frac{1}{2}[\mathbf{B}, \mathbf{r}]$.

§ 18 Fundamental conservation laws for a closed system of particles

Noether's theorem makes it possible to obtain integrals of motion if the invariance of the Lagrangian function (or, in a more general case, if the invariance of the action) with respect to some set of transformations has been already found. However, it does not explain how this kind of invariance can be found. In some cases, this invariance turns out to be related to very general assumptions about the properties of the real world.

The laws of conservation of total momentum, angular momentum, and energy of a closed system of particles are already known from a previous course of general physics. Their proof is based on (i) Newton's second and third laws and (ii) the assumption that the forces between a pair of interacting particles depend only on the difference between their radius vectors.

The Lagrangian approach, on the other hand, through the use of Noether's theorem, makes it possible to establish the fundamental connection of these conservation laws with the basic properties of space and time, such as homogeneity and isotropy space and homogeneity of time.

The assumption of *homogeneity of space* means that, from the given initial conditions, a motion of a closed system of N particles is independent on the location of the given system. This implies that the Lagrangian function of the system $L(\mathbf{r}_1, \dots, \mathbf{r}_N, \mathbf{v}_1, \dots, \mathbf{v}_N, t)$ stays unchanged when all particles of the system are shifted by the same vector $\boldsymbol{\varepsilon}$, i.e. under the transformation

$$\mathbf{r}_a \to \mathbf{r}_a + \boldsymbol{\varepsilon}, \quad t \to t\,. \tag{18.1}$$

Wherein

$$\delta\mathbf{r}_a = \boldsymbol{\varepsilon}, \quad \delta t = 0,$$

and from (17.10) it follows that

$$\left(\sum_a \mathbf{p}_a\right)\boldsymbol{\varepsilon} = \text{const}\,.$$

Further, since the vector $\boldsymbol{\varepsilon}$ is arbitrary, we obtain *the law of total momentum conservation for a closed system of particles*:

$$\sum_{a=1}^{N} \mathbf{p}_a = \mathbf{const}\,. \tag{18.2}$$

Similarly, the assumption of *space isotropy* means that the relative motion of a closed system of particles stays unchanged at any spatial rotation of this system as a whole, and therefore this rotation does not change the Lagrangian function. Let the system rotate at an angle ε around an arbitrary axis defined by the unit vector $\mathbf{n}$. Wherein

$$\mathbf{r}_a \to \mathbf{r}_a + \delta\mathbf{r}_a, \quad \delta\mathbf{r}_a = \varepsilon[\mathbf{n}, \mathbf{r}_a], \quad \delta t = 0\,, \tag{18.3}$$

and from (17.10) it follows

$$\sum_a \mathbf{p}_a \delta\mathbf{r}_a = \varepsilon \sum_a \mathbf{p}_a [\mathbf{n}, \mathbf{r}_a] = \text{const}\,,$$

or

$$\varepsilon\mathbf{n}\sum_a[\mathbf{r}_a, \mathbf{p}_a] = \mathbf{const}\,.$$

Hence, due to the arbitrariness of the direction $\mathbf{n}$, we obtain *the law of total angular momentum conservation for a closed system of particles*:

$$\mathbf{M} = \sum_{a=1}^{N}[\mathbf{r}_a, \mathbf{p}_a] = \text{const}\,. \tag{18.4}$$

Finally, *homogeneity of time* assumes that the motion of a closed system of particles does not depend on the initial moment of the motion (provided that the initial state of a system is the same). Hence, the Lagrangian function of the system stays unchanged under transformation

$$\mathbf{r}_a \to \mathbf{r}_a, \quad t \to t + \varepsilon\,. \tag{18.5}$$

Wherein

$$\delta\mathbf{r}_a = 0, \quad \delta t = \varepsilon\,,$$

and from (17.10) it follows *the law of energy conservation for a closed system of particles*:

$$E = \sum_{a=1}^{N}\mathbf{p}_a\mathbf{v}_a - L = \sum_{i=1}^{3N}p_i\dot{q}_i - L = \text{const}\,. \tag{18.6}$$

Clearly, the fact that similar conclusions can be drawn in field theory when describing systems with an infinite number of degrees of freedom renders the Lagrangian approach highly valuable. In field theory, Lagrangians are constructed under the requirements of invariance with respect to both spatial shifts and rotations and temporal shifts. Such a choice not only allows one to calculate the energy, momentum, and angular momentum of fields, but actually provides their definitions.

The integrals of motion corresponding to the Galilean and Lorentz transformations are discussed in [11], § 14 and [3], problem 4.17.

§ 19 Galilean transforms

Let the coordinate axes in the reference system $K'(x', y', z')$ be parallel to the axes in the inertial frame $K(x, y, z)$, which are assumed to be motionless, and the origin of the system K' move: $\mathbf{R}_{O'} = \mathbf{R}(t)$ (Fig. 26). Coordinates relative to the moving frame K' are introduced by the relation $\mathbf{r} = \mathbf{R}(t) + \mathbf{r}'$. For definiteness, let $\mathbf{R}(0) = 0$.

If the velocity $\mathbf{V} = \dot{\mathbf{r}}$ is constant, then the system K' is also an inertial one, and the coordinates and velocities of the particle change according to Galilean transformations:

$$\mathbf{r} = \mathbf{r}' + \mathbf{V}t'\,, \quad t = t'\,, \quad \mathbf{v} = \mathbf{v}' + \mathbf{V}\,. \tag{19.1}$$

Let the Lagrangian function of a particle in the K system be equal to

$$L(\mathbf{r}, \mathbf{v}) = \frac{1}{2}m\mathbf{v}^2 - U(\mathbf{r})\,, \tag{19.2}$$

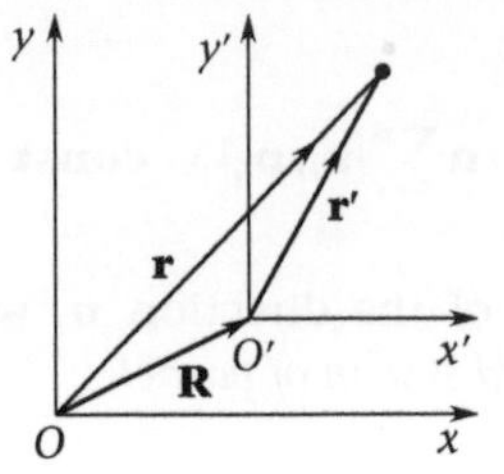

Figure 26 Two frames of reference: fixed $K(x, y, z)$ and moving $K'(x', y', z')$

then the generalized momentum and energy of this particle are:

$$\mathbf{p} = \frac{\partial L}{\partial \mathbf{v}} = m\mathbf{v}, \quad E = \mathbf{p}\mathbf{v} - L = \frac{1}{2} m\mathbf{v}^2 + U(\mathbf{r}). \tag{19.3}$$

When going to the K' system, two different ways can be proposed for obtaining the Lagrangian function. First, one can consider this transition simply as a change of variables (1). In this case, the new Lagrangian function is

$$L_1'(\mathbf{r}', \mathbf{v}', t) = L(\mathbf{r}, \mathbf{v}) = \frac{1}{2} m\,(\mathbf{v}')^2 - m\mathbf{v}'\mathbf{V} + \frac{1}{2} m\mathbf{V}^2 - U(\mathbf{r}' + \mathbf{V}t'). \tag{19.4}$$

For generalized momenta and energies in these two frames of reference we get the following relations:

$$\mathbf{p}_1' = \frac{\partial L_1'}{\partial \mathbf{v}'} = m\,(\mathbf{v}' + \mathbf{V}) = \mathbf{p}; \tag{19.5a}$$

$$E_1' = \mathbf{p}'\mathbf{v}' - L_1' = \frac{1}{2} m\,(\mathbf{v}')^2 + U - \frac{1}{2} m\mathbf{V}^2 = E - \mathbf{V}\mathbf{p}. \tag{19.5b}$$

Secondly, we can take into account that K' is an inertial frame and choose the Lagrangian function in the form of a difference between the kinetic and potential energies in the K' system:

$$L_2'(\mathbf{r}', \mathbf{v}', t) = \frac{1}{2} m\,(\mathbf{v}')^2 - U(\mathbf{r}' + \mathbf{V}t'). \tag{19.6}$$

The Lagrangian function L_2' differs from L_1' by the total derivative in time of the function

$$F(\mathbf{r}', t) = -m\mathbf{r}'\mathbf{V} + \frac{1}{2} m\mathbf{V}^2 t$$

and leads to the same Lagrangian equations as the function $L_1'(\mathbf{r}', \mathbf{v}', t)$ (see § 16). However, the transformation laws of the generalized momenta and energies have a different form than in (4):

$$\mathbf{p}_2' = \frac{\partial L_2'}{\partial \mathbf{v}'} = m\mathbf{v}' = \mathbf{p} - m\mathbf{V}, \tag{19.7a}$$

$$E_2' = \frac{1}{2} m\,(\mathbf{v}')^2 + U = E - \mathbf{V}\mathbf{p} + \frac{1}{2} m\mathbf{V}^2. \tag{19.7b}$$

The two expressions for the energy E_1' and E_2' differ by a constant.

We have already noted in § 16 that the energies E and E' can have a different time dependence. The considered transition from one inertial frame of reference to another allows us to illustrate this statement, using a simple instructive example: a motion of a ball inside a box with perfectly elastic walls. If the box is at rest in the frame K, then the velocity of the ball changes in collisions with the walls of the box only in terms of direction, but not magnitude, and the energy E is conserved. But in the system K', the ball's velocity changes upon collisions with a moving wall not only in direction, but also in magnitude, so the energy E', defined either by Eq. (5*b*) or by Eq. (7*b*), is not conserved.

§ 20 Non-inertial frames of reference

It is often convenient to use reference systems that are associated with particles moving with acceleration in the inertial frame of reference, i.e. *non-inertial frames of reference*. Such transition to the moving frame of reference is reduced to a simple change of coordinates in the Lagrangian function. In particular, the reference system associated with the Earth is a non-inertial one.

20.1 *Translational reference frame*

If the velocity $\mathbf{V}(t) = \dot{\mathbf{R}}(t)$ is not constant, then the system K' (Fig. 26) is no longer inertial. We choose the Lagrangian function in non-inertial frame K' as equal to (cf. (19.2))

$$L'(\mathbf{r}', \dot{\mathbf{r}}', t) = \frac{1}{2} m \left(\dot{\mathbf{r}}' + \mathbf{V}(t)\right)^2 - U\left(\mathbf{R}(t) + \mathbf{r}'\right). \tag{20.1}$$

Then the equations of motion in the system K' are as follows

$$m\ddot{\mathbf{r}}' = -\frac{\partial U}{\partial \mathbf{r}'} - m\mathbf{W}, \tag{20.2}$$

where $\mathbf{W} = \dot{\mathbf{V}}(t)$ is the acceleration of the system K'. The change of equations of motion under the transition to the reference frame K' is reduced to adding the *inertia force* (equal to $(-m\mathbf{W})$) to the ordinary force acting on the particle.[6]

At the space station, the inertial force acting on the astronaut (due to the accelerated motion of the station in the Earth's gravity field) is completely compensated by the force of gravity acting on the astronaut. As a consequence, weightlessness occurs.

In a reference system, whose origin is associated with the centre of the Earth and the axes oriented according to the stars, the forces with which the Sun and the Moon act are compensated by the forces of inertia for those particles that are located in the centre of the Earth. For the particles located on the surface of the Earth, there is no full compensation. The uncompensated part is *tidal forces* (see § 55).

[6] In the Newtonian mechanics framework, inertia forces are different from ordinary forces, which results from the interaction of particles, as the forces of inertia arise without interaction when switching to a non-inertial frame of reference. Therefore, it is impossible to apply Newton's third law ('action equals resistance') to them. However, this does not affect the solving of the problem of the particle motion.

20.2 *Rotating reference frame*

Let the reference frame $K'(x', y', z')$ rotate with an angular velocity $\boldsymbol{\Omega} = \boldsymbol{\Omega}(t)$ relative to the inertial frame $K(x, y, z)$ and let their origins coincide. It is clear that the radius vector $\mathbf{r}$ which specifies the position of the material point in the K frame coincides with the radius vector $\mathbf{r}'$ which defines the position of the same point in the K' system, although the components of these vectors x, y, z and x', y', z', generally speaking, do not coincide.

If a particle is at rest in the K' frame, then in the K frame its velocity is

$$\mathbf{v} = [\boldsymbol{\Omega}, \mathbf{r}] = [\boldsymbol{\Omega}, \mathbf{r}']$$

(cf. (18.3)). If, however, in the K' frame the particle moves with the velocity $\mathbf{v}'$, then by adding this value to $[\boldsymbol{\Omega}, \mathbf{r}']$, we finally obtain

$$\mathbf{r} = \mathbf{r}', \quad \mathbf{v} = \mathbf{v}' + [\boldsymbol{\Omega}, \mathbf{r}']. \tag{20.3}$$

The Lagrangian function $L'(\mathbf{r}', \mathbf{v}', t)$ for a particle in a rotating reference system is obtained from the Lagrangian function (19.2) by the change of variables (3):

$$L'(\mathbf{r}', \mathbf{v}', t) = \frac{1}{2} m \left(\mathbf{v}'\right)^2 + m\mathbf{v}' [\boldsymbol{\Omega}, \mathbf{r}'] + \frac{1}{2} m [\boldsymbol{\Omega}, \mathbf{r}']^2 - U(\mathbf{r}'). \tag{20.4}$$

From here we get both the generalized momentum and the angular momentum:

$$\mathbf{p}' = \frac{\partial L'}{\partial \mathbf{v}'} = m \left(\mathbf{v}' + [\boldsymbol{\Omega}, \mathbf{r}']\right), \quad \mathbf{M}' = \mathbf{r}' \times \mathbf{p}'. \tag{20.5}$$

Taking into account Eqs. (3) and (5), we find the relationship between $\mathbf{p}'$ and $\mathbf{M}'$ with analogous quantities in the K system:

$$\mathbf{p}' = \mathbf{p}, \quad \mathbf{M}' = \mathbf{M}. \tag{20.6}$$

For the energy E' in the K' frame, we have

$$E' = \mathbf{p}'\mathbf{v}' - L' = \frac{1}{2} m \left(\mathbf{v}'\right)^2 + U(\mathbf{r}') - \frac{1}{2} m [\boldsymbol{\Omega}, \mathbf{r}']^2. \tag{20.7}$$

In addition to kinetic and potential energies, E' contains the term

$$-\frac{1}{2} m [\boldsymbol{\Omega}, \mathbf{r}']^2,$$

called the *centrifugal energy*. We find the relationship of E' with energy E in system K by using (3) and (5):

$$E = \mathbf{p}\mathbf{v} - L = \mathbf{p}'(\mathbf{v}' + [\boldsymbol{\Omega}, \mathbf{r}']) - L' = E' + \boldsymbol{\Omega}\mathbf{M}'. \tag{20.8}$$

The equation of motion in a rotating system is the Lagrangian equation for the new coordinates

$$\frac{d}{dt} \frac{\partial L'}{\partial \mathbf{v}'} = \frac{\partial L'}{\partial \mathbf{r}'}.$$

After a simple rearrangement, this equation has the form

$$m\frac{d\mathbf{v}'}{dt} = -\frac{\partial U}{\partial \mathbf{r}'} + 2m[\mathbf{v}', \boldsymbol{\Omega}] + m[\boldsymbol{\Omega}, [\mathbf{r}', \boldsymbol{\Omega}]] + m[\mathbf{r}', \dot{\boldsymbol{\Omega}}]\,. \tag{20.9}$$

In addition to the usual potential force $-\partial U/\partial \mathbf{r}'$, the right side of this equation contains forces of inertia due to the non-inertiality of the reference system. Those are the *Coriolis force*

$$2m[\mathbf{v}', \boldsymbol{\Omega}]\,,$$

the *centrifugal force of inertia*

$$m[\boldsymbol{\Omega}, [\mathbf{r}', \boldsymbol{\Omega}]]$$

and the *inertia force*

$$m\,[\mathbf{r}', \dot{\boldsymbol{\Omega}}]\,,$$

due to the uneven rotation.

In the reference frame associated with the Earth, the forces of inertia manifest themselves in such large-scale phenomena as deviations of the wind directions (right-ward in the north and left-ward in the southern hemispheres) and the major oceanic and river currents.

For problems where such a transition to a rotating frame of reference is most useful, see 9.23, 9.25, and 9.28 from [3], .

20.3 *Larmor's theorem*

The transition to a rotating frame of reference turns out to be very effective in the next problem. Consider a finite motion of a charged particle in the potential field $U(r)$ where additionally there exists a uniform constant magnetic field $\mathbf{B}$. What will the motion of the particle look like in this case? If the magnetic field is small, it is not difficult to obtain a general result known as the *Larmor's theorem.*

Let us write the Lagrangian function of the considered problem L in the inertial reference frame K:

$$L = \frac{1}{2}m\mathbf{v}^2 - U(r) + \frac{e}{c}\mathbf{v}\mathbf{A}(\mathbf{r}), \tag{20.10}$$

where e is the charge of the particle and $\mathbf{A}(\mathbf{r})$ is the vector potential, which can be chosen in the form

$$\mathbf{A}(\mathbf{r}) = \frac{1}{2}\,[\mathbf{B}, \mathbf{r}]. \tag{20.11}$$

Let us now consider the same motion in the reference frame K', rotating with a constant angular velocity $\boldsymbol{\Omega}$ with respect to the K system. Relations between the radius vectors $\mathbf{r}$ and $\mathbf{r}'$ and the velocities $\mathbf{v}$ and $\mathbf{v}'$ in the reference frames K and K' are given by Eqs. (3).

Substituting these expressions into (10), we obtain the Lagrangian of our problem L' in the reference frame K':

$$L' = \frac{1}{2} m \left(\mathbf{v}' + [\boldsymbol{\Omega}, \mathbf{r}']\right)^2 - U(r') + \frac{e}{2c} \left(\mathbf{v}' + [\boldsymbol{\Omega}, \mathbf{r}']\right) \cdot [\mathbf{B}, \mathbf{r}'] .$$

It can be represented as

$$L' = L'_0 + \delta L' , \; L'_0 = \frac{1}{2} m(\mathbf{v}')^2 - U(r') , \tag{20.12}$$

$$\delta L' = m\mathbf{v}'[\boldsymbol{\Omega}, \mathbf{r}'] + \frac{e}{2c}\mathbf{v}'[\mathbf{B}, \mathbf{r}'] + \frac{1}{2} m[\boldsymbol{\Omega}, \mathbf{r}']^2 + \frac{e}{2c}[\boldsymbol{\Omega}, \mathbf{r}'] \cdot [\mathbf{B}, \mathbf{r}'].$$

If we choose for $\boldsymbol{\Omega}$ the value (called the *Larmor's frequency*)

$$\boldsymbol{\Omega}_{\mathrm{L}} = -\frac{e\mathbf{B}}{2mc}, \tag{20.13}$$

then the quantities of the first order in the field B totally cancel each other and $\delta L'$ turns out to be a second-order quantity:

$$\delta L' = -\frac{e^2}{8mc^2} \left[\mathbf{B}, \mathbf{r}'\right]^2 . \tag{20.14}$$

This shows that if a magnetic field is sufficiently small, then for a finite motion of the particle the term $\delta L'$ is also small and therefore can be neglected. Thus, the motion of a particle in the reference frame K' is determined only by the potential field $U(r')$.

As an example, we consider the motion of a charged particle in the Coulomb field $U(r) = -\alpha/r$ and a small uniform and constant magnetic field $\mathbf{B}$ at energy $E<0$. When we go to the reference frame K', the particle's orbit becomes an ordinary ellipse. And all the influence of a weak magnetic field in the initial reference frame K is reduced to the precession of this ellipse about the direction of the field $\mathbf{B}$ with the Larmor's frequency (13). In this case, the angular momentum of the particle also precesses around the direction of the magnetic field.

§ 21 Deviation of a freely falling body from the vertical

The reference frame associated with the Earth is not an inertial one. That is why in the northern hemisphere there occurs a east- and southward *deviation of a body's trajectory from the vertical line*. By definition, *a vertical is the position of a resting plumb line*. In solving this problem, we will consider the Earth as a homogeneous ball of radius $R = 6\,400$ km rotating with a constant angular velocity $\boldsymbol{\Omega}$. Besides, we assume that the centres of the non-inertial frame and the centre of the Earth coincide, and that coordinates of a body in this frame will be denoted as $\mathbf{R} + \mathbf{r}$. We will assume that the height of the body above the Earth h is small compared to R. Similarly, we will assume that the centrifugal acceleration $\sim R\Omega^2$ is small compared to the acceleration of free fall on the surface of the Earth g.

Thus, there are two small parameters in the problem:

$$\epsilon_1 = \frac{h}{R} \lesssim \epsilon_2 = \frac{R\Omega^2}{g} \approx 0.0034.$$

In the equation of motion (G is the gravitational constant, M the Earth mass)

$$\ddot{\mathbf{r}} = -GM\frac{\mathbf{R}+\mathbf{r}}{|\mathbf{R}+\mathbf{r}|^3} + 2[\mathbf{v}, \boldsymbol{\Omega}] + [\boldsymbol{\Omega}, [\mathbf{R}+\mathbf{r}, \boldsymbol{\Omega}]] \tag{21.1}$$

we expand the first term on the right-hand side into a series in the small parameter $r/R \lesssim \epsilon_1$:

$$\ddot{\mathbf{r}} = \mathbf{g} + 2[\mathbf{v}, \boldsymbol{\Omega}] + \mathbf{g}_1 + [\boldsymbol{\Omega}, [\mathbf{r}, \boldsymbol{\Omega}]], \tag{21.2}$$

where we use the notation

$$\mathbf{g} = -GM\frac{\mathbf{R}}{R^3} + [\boldsymbol{\Omega}, [\mathbf{R}, \boldsymbol{\Omega}]], \; ; \; \mathbf{g}_1 = \frac{GM}{R^2}\left(3\mathbf{R}\frac{(\mathbf{rR})}{R^3} - \frac{\mathbf{r}}{R}\right).$$

The Coriolis acceleration $|2[\mathbf{v}, \boldsymbol{\Omega}]| \sim gt\Omega \sim \sqrt{\epsilon_1\epsilon_2}\, g$ and $g_1 \sim \epsilon_1 g$ has the first order of smallness, while the acceleration $|[\boldsymbol{\Omega}, [\mathbf{r}, \boldsymbol{\Omega}]]| \sim \epsilon_1\epsilon_2 g$, has the second. Also, note that the gravitational acceleration $\mathbf{g}$ does not coincide directionally with the vector $(-\mathbf{R})$ due to the small centrifugal acceleration $[\boldsymbol{\Omega}, [\mathbf{R}, \boldsymbol{\Omega}]]$.

The initial data of our problem are as follows:

$$\mathbf{r}(0) = \mathbf{h}, \quad \mathbf{v}(0) = 0.$$

It is important to take into account that the vertical $\mathbf{h}$ is antiparallel to the vector $\mathbf{g}$ and forms a small angle $\alpha = \epsilon_2 \sin\lambda\cos\lambda$ with the direction of vector $\mathbf{R}$. Here λ is the north (geocentric) latitude, that is, the angle between the equatorial plane and the vector $\mathbf{R}$. If we choose the z-axis along the vertical, the x-axis along a meridian to the south, and the y-axis along latitude to the east, then

$$\mathbf{g} = (0, 0, -g), \; \mathbf{R} = R(\sin\alpha, 0, \cos\alpha), \quad \boldsymbol{\Omega} = \Omega(-\cos(\alpha+\lambda), 0, \sin(\alpha+\lambda)).$$

In zero approximation, the particle moves with the acceleration $-g$ along the z-axis. In the first approximation, the Coriolis acceleration leads only to the eastward deviation, and the amount of displacement is $y \sim \sqrt{\epsilon_1\epsilon_2}\, gt^2 \sim \sqrt{\epsilon_1\epsilon_2}\, h$. The acceleration $\mathbf{g}_1$ has $\sim \epsilon_1 g$ components along the z-axis, while along the x- and y-axes there are only second-order components. Therefore, in the first approach, the influence of $\mathbf{g}_1$ will only lead to an increase in the time of falling from height h by $\sim \epsilon_1\sqrt{2h/g}$. The southward deviation occurs only in the second order.

Let us write equation (2) in projections onto the chosen axes, holding the terms of zero, first, and second orders of smallness for the components z, y, and x, respectively:

$$\ddot{z} = -g, \quad \ddot{y} = -2\dot{z}\Omega\cos\lambda,$$

$$\ddot{x} = 2\dot{y}\Omega\sin\lambda + \frac{g}{R}(3z\sin\alpha - x) + \Omega^2 z\sin\lambda\cos\lambda.$$

Solving these equations via the method of successive approximations, we find

$$z = h - \frac{1}{2}gt^2, \quad y = \frac{1}{3}gt^3\Omega\cos\lambda, \quad x = 2ht^2\Omega^2\sin\lambda\cos\lambda.$$

Substituting the fall time $t = \sqrt{2h/g}$, we find the following deviations: (i) eastward

$$y = \frac{1}{3}\sqrt{8\epsilon_1\epsilon_2}\, h\cos\lambda$$

and (ii) southward

$$x = 2\epsilon_1\epsilon_2 h\sin 2\lambda.$$

At a height of 100 m, the eastward deviation at the latitude of Novosibirsk ($\lambda = 55°$) is 12 mm, and the southward one is 0.01 mm.

§ 22 Effective Lagrangian function for electromechanical systems

Using ideal holonomic constraints is not the only possibility to decrease the number of considered coordinates while preserving the Lagrangian form of equations. Consider the following, less obvious, example. Motion of a liquid column in a vertical *U*-shaped tube (Fig. 27) can be investigated based on the Lagrangian function

$$L = \frac{1}{2}m\dot{x}^2 - \frac{mgx^2}{l}, \tag{22.1}$$

where m is its mass, l is the length of a liquid column, and x is the height level in one column, counted from the equilibrium position. In this approach, we neglect such factors as friction, deviation of the liquid surface from the flat form, etc.

As another example, we consider small oscillations of a particle of mass M suspended on a spring of the elasticity coefficient k and mass m in the gravitational field. If we neglect the mass of the spring, then the frequency of small oscillations is $\omega_0 = \sqrt{k/M}$. Assuming that the spring is homogeneous, we find the correction to the oscillation frequency due to the small mass of the spring, $m \ll M$. The spring is a system with an infinite number of degrees of freedom. However, in the considered motion, it experiences only such compression and stretching which can be described by specifying just one quantity, namely, the length of the spring. To study such a motion, the Lagrangian functions with one degree of freedom

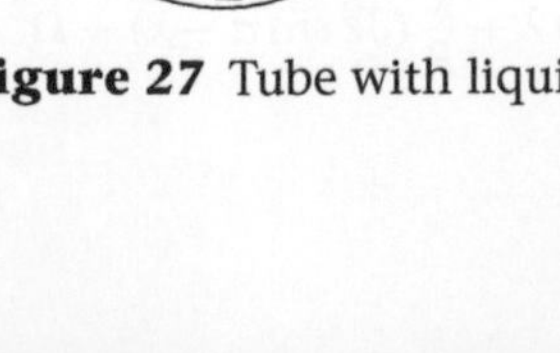

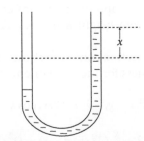

Figure 27 Tube with liquid

will suffice. By choosing the deviation x of the particle from the equilibrium position as a coordinate, we add to the 'ordinary' Lagrangian function

$$L_0 = \frac{1}{2}\left(M\dot{x}^2 - kx^2\right)$$

the spring's kinetic energy

$$T = \frac{1}{2}\rho\int_0^l v^2(\xi)d\xi.$$

Here $\rho = m/l$ is the linear density of the spring, $l(t)$ is its length, and $v(\xi) = \dot{x}\xi/l$ is a velocity of a spring's point, currently located at a distance ξ from the suspension point, thus $T = m\dot{x}^2/6$. As a result, one-third of the spring mass $m/3$ is added to the particle mass M in the Lagrangian function, and the oscillation frequency turns out to be equal to

$$\omega = \sqrt{\frac{k}{M+(m/3)}} \approx \sqrt{\frac{k}{M}}\left(1 - \frac{m}{6M}\right). \tag{22.2}$$

In these simple examples, we did not prove that the motions are described by a small number of coordinates, but rather assumed that as obvious, based on 'physical considerations'. In fact, we do the same in more complex cases, assuming, for example, that constraints are ideal. Here it would perhaps be appropriate to recall that all the classical mechanics objects, such as a material point, a rigid body, ideal connections and etc., are the result of idealization.
Therefore, the results of any calculations are approximate.

Consider now an electrical circuit consisting of a capacitor of capacitance C and a solenoid of inductance $\mathcal{L}$ (Fig. 28). Let $q(t)$ be the charge on the top plate of capacitor, then the current in the solenoid is $\dot{q}$. Neglecting losses due to resistance and radiation, we obtain, as the Kirchhoff's equation, the equation of oscillations in the circuit (in the SI system of units):

$$\mathcal{L}\ddot{q} + \frac{q}{C} = 0. \tag{22.3}$$

At the same time, this equation can also be the Lagrangian equation if we use the Lagrangian function

$$L(q,\dot{q}) = \frac{1}{2}\mathcal{L}\dot{q}^2 - \frac{q^2}{2C} \tag{22}$$

with a generalized coordinate q equal to the charge on the capacitor plate. The energy of magnetic field in the solenoid acts as kinetic energy, while the energy of the electric field

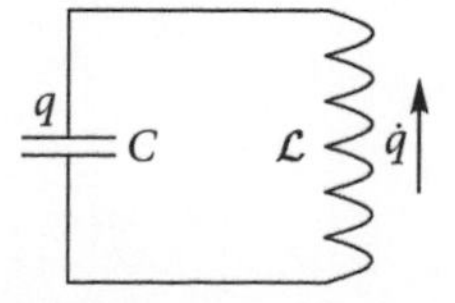

Figure 28 Electrical circuit

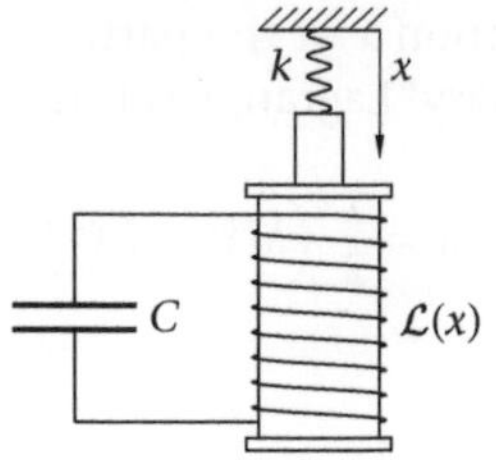

Figure 29 Electromechanical system

in the capacitor appears as the potential energy. Note that Lagrangian (4) gives the correct value of the energy of the system

$$E = \frac{1}{2}\mathcal{L}\dot{q}^2 + \frac{q^2}{2C}. \tag{22.4}$$

The electromagnetic fields considered in this section form a continuous system. They are described by Maxwell's equations. For a continuous medium, the latter can also be represented in the form of the Lagrangian equations. When we go to electric circuits with concentrated parameters, we actually describe the electric field in a capacitor by only one generalized coordinate, namely the charge q in capacitor, and the magnetic field in the solenoid by the current $\dot{q}$. At the same time, we neglect the possibility of 'excitation' of other degrees of freedom of the electromagnetic field such as, for example, electromagnetic waves. Transition from the Lagrangian of the electromagnetic field to the Lagrangian of the form (4) is demonstrated, for instance, in [3], problem 4.24. Such a description of a continuous system using a small number of 'essential' generalized coordinates is similar to the description of motion of a liquid column in the system Fig. 27. It is wonderful that such an approach can also be used to describe systems for whose motion both mechanical and electrodynamic degrees of freedom are important.

As an example, we consider the system Fig. 29 consisting of the $\mathcal{L}C$-circuit and a body which is a solenoid core suspended on a spring, so the inductance of the solenoid depends on the displacement of the body. The Lagrangian function of this *electromechanical system* with two degrees of freedom

$$L(x, q, \dot{x}, \dot{q}) = \frac{1}{2}\left[m\dot{x}^2 + \mathcal{L}(x)\dot{q}^2 - \mathcal{C}q^2 - kx^2\right] \tag{22.5}$$

leads to equations of motion connecting the charge of the capacitor q with the displacement of the solenoid core x, which is counted from its equilibrium position in the absence of current. For other examples of electromechanical systems, see [3], problems 4.26, 4.27.

CHAPTER III

Oscillations

§ 23 Linear oscillations

Linear oscillations are encountered in almost every area of physics. In this section, we introduce useful notations and recall the main definitions, using a one-dimensional system as an example.

23.1 *One degree of freedom*

Let the Lagrangian of a system with one degree of freedom have the form

$$L(q,\dot q) = T(q,\dot q) - U(q)\,, \quad T(q,\dot q) = \frac{1}{2}\,a(q)\,\dot q^2 \ge 0\,. \tag{23.1}$$

If q_0 is the point of minimum potential energy $U(q)$, then a small excess of total energy over the minimum of the potential energy will result in periodic oscillations of small amplitude. In this case, we can expand potential energy $U(q)$ in a series in small deviation $x = q - q_0$:

$$U(q) = U(q_0) + \frac{1}{2}\,kx^2\,, \quad \left.\frac{dU}{dq}\right|_{q_0} = 0\,, \quad \left.\frac{d^2U}{dq^2}\right|_{q_0} = k > 0. \tag{23.2}$$

(Note that the case $k=0$ corresponds to non-linear oscillations.) We also expand the function $a(q)$ near q_0:

$$a(q) = m + \mathcal{O}(x)\,,\ m = a(q_0)\,. \tag{23.3}$$

When restricting ourselves to terms of second order in x and $\dot x = \dot q$, we get (discarding the constant $U(q_0)$)

$$L(x,\dot x) = \frac{1}{2}\,m\dot x^2 - \frac{1}{2}\,kx^2\,. \tag{23.4}$$

The Lagrangian equation

$$m\ddot x + kx = 0 \tag{23.5}$$

under substitution

$$x = A\cos(\omega t + \varphi) \tag{23.6}$$

Lectures on Analytical Mechanics. G. L. Kotkin, V. G. Serbo, A. I. Chernykh, Oxford University Press. © G. L. Kotkin, V. G. Serbo, A. I. Chernykh (2024). DOI: 10.1093/oso/9780198894674.003.0003

is reduced to the algebraic equation

$$-m\omega^2 + k = 0\,, \tag{23.7}$$

so it follows

$$\omega = \sqrt{\frac{k}{m}}\,. \tag{23.8}$$

The *amplitude* of the oscillations is denoted by A, $\omega t + \varphi$ is the *phase*, φ is the *initial phase*, and ω is the *(circular) frequency*. The period of oscillations is

$$T = \frac{2\pi}{\omega}\,. \tag{23.9}$$

The quantity $\omega^2 = k/m$ is positive since both m and k are positive. Of course, the transition from the original mechanical system (1) to the linearized one (4) is valid only for sufficiently small x (or A).

23.2 *Oscillations of systems with many degrees of freedom*

Let us consider now the case of several degrees of freedom:

$$L = T - U(q_1, q_2, \ldots, q_s)\,, \quad T = \frac{1}{2}\sum_{ij=1}^{s} a_{ij}(q)\dot{q}_i\dot{q}_j \geq 0\,. \tag{23.10}$$

Given that the product $\dot{q}_i\dot{q}_j$ is symmetric under the replacement of $i \leftrightarrow j$, it is always possible to select a symmetric matrix $a_{ij}(q)$:

$$a_{ij}(q) = a_{ji}(q)\,. \tag{23.11}$$

Let q_{i0} (with $i = 1, 2, \ldots, s$) be the point of minimum potential energy. Assuming small oscillation amplitudes, the potential energy can be expanded in terms of a small deviation series $x_i = q_i - q_{i0}$ as is done for the one-dimensional case, that is:

$$U(q) = \frac{1}{2}\sum_{ij} k_{ij}x_ix_j + \text{const}\,, \quad k_{ij} = \left.\frac{\partial^2 U}{\partial q_i \partial q_j}\right|_{q_{i0}} = k_{ji}\,. \tag{23.12}$$

Since at $x_i = 0$ the potential energy has a minimum, the quadratic form (12) is a positive definite function:

$$\sum_{ij} k_{ij}x_ix_j \geq 0\,. \tag{23.13}$$

We also expand the function $a_{ij}(q)$ near q_{i0}:

$$a_{ij}(q) = m_{ij} + \mathcal{O}(x_k); \quad m_{ij} = a_{ij}(q_{k0}) = m_{ji}\,. \tag{23.14}$$

Since the kinetic energy is positive, the quadratic form

$$\sum_{ij} m_{ij}\dot{x}_i\dot{x}_j \geq 0 \tag{23.15}$$

is also a positive definite function. Restricting ourselves to terms of the second order in x_i and $\dot{x}_i = \dot{q}_i$, we get

$$L = \frac{1}{2} \sum_{ij} \left(m_{ij}\dot{x}_i\dot{x}_j - k_{ij}x_ix_j \right) . \tag{23.16}$$

In what follows, it is convenient to switch to vector notation. We introduce a displacement vector[1]

$$\mathbf{x} = \begin{pmatrix} x_1 \\ x_2 \\ \vdots \\ x_s \end{pmatrix}$$

and the scalar product of vectors:

$$(\mathbf{x}, \mathbf{y}) = \sum_{i=1}^{s} x_i y_i .$$

We also introduce the mass and stiffness matrices

$$\hat{m} = \begin{pmatrix} m_{11} & m_{12} & \dots & m_{1s} \\ m_{21} & m_{22} & \dots & m_{2s} \\ \dots & \dots & \dots & \dots \\ m_{s1} & m_{s2} & \dots & m_{ss} \end{pmatrix}, \quad \hat{k} = \begin{pmatrix} k_{11} & k_{12} & \dots & k_{1s} \\ k_{21} & k_{22} & \dots & k_{2s} \\ \dots & \dots & \dots & \dots \\ k_{s1} & k_{s2} & \dots & k_{ss} \end{pmatrix}.$$

In these notations, the Lagrangian takes the form

$$L(\mathbf{x}, \dot{\mathbf{x}}) = \frac{1}{2} (\dot{\mathbf{x}}, \hat{m}\,\dot{\mathbf{x}}) - \frac{1}{2} (\mathbf{x}, \hat{k}\,\mathbf{x}) , \tag{23.17}$$

in which

$$(\dot{\mathbf{x}}, \hat{m}\,\dot{\mathbf{x}}) \geq 0 , \quad (\mathbf{x}, \hat{k}\,\mathbf{x}) \geq 0, \quad \hat{m}^T = \hat{m}, \quad \hat{k}^T = \hat{k} , \tag{23.18}$$

where $\hat{m}^T$ denotes the matrix transposed with respect to the matrix $\hat{m}$. The Lagrangian equations

$$\frac{d}{dt}\frac{\partial L}{\partial \dot{x}_l} = \frac{\partial L}{\partial x_l}$$

in these notations

$$\frac{d}{dt}\frac{\partial L}{\partial \dot{\mathbf{x}}} = \frac{\partial L}{\partial \mathbf{x}}$$

take a form similar to (5):

$$\hat{m}\ddot{\mathbf{x}} + \hat{k}\mathbf{x} = 0 . \tag{23.19}$$

[1] The components of this vector, as well as the components of the mass and stiffness matrices, can have different dimensions.

Substitution

$$\mathbf{x} = \mathbf{A}\cos(\omega t + \varphi) \tag{23.20}$$

reduces these equations to a system of algebraic linear homogeneous equations

$$\left(-\omega^2 \hat{m} + \hat{k}\right) \mathbf{A} = 0. \tag{23.21}$$

This system has a non-trivial solution only if its determinant equals zero:

$$\left|-\omega^2 \hat{m} + \hat{k}\right| = 0. \tag{23.22}$$

Let $\omega_1^2, \omega_2^2, \ldots, \omega_s^2$ be the roots of this equation (some of these roots may coincide; this case will be discussed in the next section). Quantities ω_α are called *eigenfrequencies*. Substituting one of these roots ω_α^2 into equation (21), we obtain an equation for determining the components of the corresponding vector $\mathbf{A}^{(\alpha)}$:

$$\left(-\omega_\alpha^2 \hat{m} + \hat{k}\right) \mathbf{A}^{(\alpha)} = 0. \tag{23.23}$$

Of course, if $\mathbf{A}^{(\alpha)}$ is a solution of this homogeneous equation, then $a\mathbf{A}^{(\alpha)}$, where a is an arbitrary number, is also a solution of this equation. As a result, each root ω_α^2 (or each frequency ω_α) corresponds to an oscillation

$$\mathbf{x}^{(\alpha)}(t) = \mathbf{A}^{(\alpha)} Q_\alpha(t), \quad Q_\alpha(t) = a_\alpha \cos(\omega_\alpha t + \varphi_\alpha), \tag{23.24}$$

at which all particles move with the same frequency (and in the same phase or the antiphase). Such a motion is called *a normal oscillation or mode*. The complete solution is the sum of particular solutions

$$\mathbf{x} = \sum_{\alpha=1}^{s} \mathbf{A}^{(\alpha)} Q_\alpha(t). \tag{23.25}$$

This solution can also be written in the notation for the individual components of the vector

$$x_i = \sum_{\alpha=1}^{s} A_i^{(\alpha)} Q_\alpha(t). \tag{23.26}$$

This complete solution contains s arbitrary amplitudes a_α and phases φ_α, which can be determined by setting the initial coordinates $\mathbf{x}(0)$ and velocities $\dot{\mathbf{x}}(0)$. Thus, solution of small amplitude oscillations, described by linear differential equations, can be obtained in a general form.

23.3 *Flat double pendulum*

Here we consider small oscillations of a flat double pendulum (see Fig. 23). The Lagrangian of this system was found in [1], § 5, problem 1, and for small angles $|\varphi_i| \ll 1$ and $l_1 = l_2 = l$ takes the form

$$L = \frac{1}{2} ml^2(8\dot{\varphi}_1^2 + 4\dot{\varphi}_1\dot{\varphi}_2 + \dot{\varphi}_2^2) - \frac{1}{2} mgl(4\varphi_1^2 + \varphi_2^2).$$

The mass and stiffness matrices are given by

$$\hat{m} = ml^2 \begin{pmatrix} 8 & 2 \\ 2 & 1 \end{pmatrix}, \quad \hat{k} = mgl \begin{pmatrix} 4 & 0 \\ 0 & 1 \end{pmatrix}.$$

We are looking for the solution of the equations of motion

$$8\ddot{\varphi}_1 + 2\ddot{\varphi}_2 + 4\omega_0^2\varphi_1 = 0, \quad 2\ddot{\varphi}_1 + \ddot{\varphi}_2 + \omega_0^2\varphi_2 = 0, \quad \omega_0 = \sqrt{g/l}$$

in the form

$$\mathbf{x} = \begin{pmatrix} \varphi_1 \\ \varphi_2 \end{pmatrix} = \begin{pmatrix} A_1 \\ A_2 \end{pmatrix} \cos(\omega t + \chi)$$

and get the system of equations for the coefficients A_i:

$$(-8\omega^2 + 4\omega_0^2)A_1 - 2\omega^2 A_2 = 0, \quad -2\omega^2 A_1 + (-\omega^2 + \omega_0^2)A_2 = 0.$$

Setting the determinant of the above system of linear equations to zero, we obtain

$$\omega^4 - 3\omega^2\omega_0^2 + \omega_0^4 = 0$$

which provides us with the eigenfrequencies

$$\omega_{1,2} = \sqrt{\frac{3 \mp \sqrt{5}}{2}}\, \omega_0$$

and their corresponding normal oscillations

$$\mathbf{x}^{(1)} = \mathbf{A}^{(1)} Q_1(t), \quad \mathbf{x}^{(2)} = \mathbf{A}^{(2)} Q_2(t); \quad Q_\alpha(t) = a_\alpha \cos(\omega_\alpha t + \chi_\alpha).$$

The vectors

$$\mathbf{A}^{(1)} = \begin{pmatrix} 1 \\ \sqrt{5} - 1 \end{pmatrix}, \quad \mathbf{A}^{(2)} = \begin{pmatrix} -1 \\ \sqrt{5} + 1 \end{pmatrix}$$

set the direction of the new coordinate axes Q_1 and Q_2 in the plane φ_1, φ_2.

The oscillations along each of these directions are performed with one frequency, i.e. with ω_1 along $\mathbf{A}^{(1)}$ and with ω_2 along $\mathbf{A}^{(2)}$ (Fig. 30). Contrarily, the time dependence of each coordinate can be expressed as the sum of oscillations with two distinct eigenfrequencies:

$$\varphi_1 = Q_1 - Q_2,$$
$$\varphi_2 = (\sqrt{5} - 1)Q_1 + (\sqrt{5} - 1)\, Q_2.$$

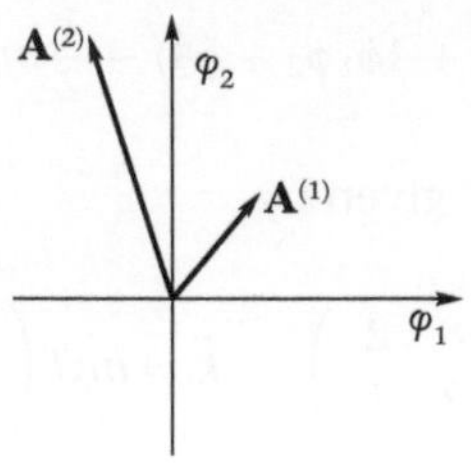

Figure 30 Vectors of normal oscillations of a double flat pendulum shown in Fig. 23

The angle between the vectors $\mathbf{A}^{(1)}$ and $\mathbf{A}^{(2)}$ (i.e., between axes Q_1 and Q_2) is not equal to 90°; indeed,

$$\left(\mathbf{A}^{(1)}, \mathbf{A}^{(2)}\right) = 3 \neq 0 .$$

However, it is easy to verify that these vectors satisfy the following relations:

$$\left(\mathbf{A}^{(1)}, \hat{k}\mathbf{A}^{(2)}\right) = \left(\mathbf{A}^{(1)}, \hat{m}\mathbf{A}^{(2)}\right) = 0, \tag{23.27}$$

for example,

$$\left(\mathbf{A}^{(1)}, \frac{\hat{k}}{mgl}\mathbf{A}^{(2)}\right) = (1, \sqrt{5}-1)\begin{pmatrix} 4 & 0 \\ 0 & 1 \end{pmatrix}\begin{pmatrix} -1 \\ \sqrt{5}+1 \end{pmatrix} =$$
$$= (1, \sqrt{5}-1)\begin{pmatrix} -4 \\ \sqrt{5}+1 \end{pmatrix} = 0 .$$

Below in § 24, we will show that the relations (27) found in this example are also valid for the general case.

§ 24 Orthogonality of normal oscillations. The case of frequency degeneracy

24.1 *Orthogonality of normal oscillations*

Let ω_α and ω_β be different eigenfrequencies: $\omega_\alpha \neq \omega_\beta$. The corresponding eigenvectors $\mathbf{A}^{(\alpha)}$ and $\mathbf{A}^{(\beta)}$ satisfy the equations

$$\omega_\alpha^2 \hat{m}\mathbf{A}^{(\alpha)} = \hat{k}\mathbf{A}^{(\alpha)}, \quad \omega_\beta^2 \hat{m}\mathbf{A}^{(\beta)} = \hat{k}\mathbf{A}^{(\beta)} . \tag{24.1}$$

We multiply the second equation by $\mathbf{A}^{(\alpha)}$

$$\omega_\beta^2\left(\mathbf{A}^{(\alpha)}, \hat{m}\mathbf{A}^{(\beta)}\right) = \left(\mathbf{A}^{(\alpha)}, \hat{k}\mathbf{A}^{(\beta)}\right) \tag{24.2}$$

and subtract it from the first equation multiplied by $\mathbf{A}^{(\beta)}$:

$$\omega_\alpha^2\left(\mathbf{A}^{(\beta)}, \hat{m}\mathbf{A}^{(\alpha)}\right) - \omega_\beta^2\left(\mathbf{A}^{(\alpha)}, \hat{m}\mathbf{A}^{(\beta)}\right) =$$

$$= \left(\mathbf{A}^{(\beta)}, \hat{k}\mathbf{A}^{(\alpha)}\right) - \left(\mathbf{A}^{(\alpha)}, \hat{k}\mathbf{A}^{(\beta)}\right). \tag{24.3}$$

Taking into account that for any real matrix $\hat{n}$ the relation

$$\left(\mathbf{A}, \hat{n}\mathbf{B}\right) = \left(\hat{n}^T \mathbf{A}, \mathbf{B}\right)$$

holds and the fact that the matrices $\hat{k}$ and $\hat{m}$ are symmetric (see (23.18)), we obtain from (3)

$$\left(\omega_\alpha^2 - \omega_\beta^2\right)\left(\mathbf{A}^{(\alpha)}, \hat{m}\,\mathbf{A}^{(\beta)}\right) = 0. \tag{24.4}$$

Since $\omega_\alpha^2 - \omega_\beta^2 \neq 0$, this implies relation

$$\left(\mathbf{A}^{(\alpha)}, \hat{m}\,\mathbf{A}^{(\beta)}\right) = 0. \tag{24.5a}$$

Moreover, taking into account (2), we obtain the further relation

$$\left(\mathbf{A}^{(\alpha)}, \hat{k}\,\mathbf{A}^{(\beta)}\right) = 0. \tag{24.5b}$$

The relations obtained mean that the oscillations of $\mathbf{x}^{(\alpha)}(t) = \mathbf{A}^{(\alpha)} Q_\alpha(t)$ and $\mathbf{x}^{(\beta)}(t) = \mathbf{A}^{(\beta)} Q_\beta(t)$, corresponding to different frequencies, are mutually orthogonal if their scalar product is defined using metric tensors m_{ij} or k_{ij} (we say that, $\mathbf{x}^{(\alpha)}$ and $\mathbf{x}^{(\beta)}$ are orthogonal in the 'metric of mass' or the 'metric of stiffness').

24.2 *The case of frequency degeneracy. Normal coordinates*

Consider the case where the roots of the characteristic equation (23.22) contain multiple roots; this is also known as the *degenerate frequencies* case. Let us assume that two distinct solutions $\mathbf{x}^{(1)}$ and $\mathbf{x}^{(2)}$ correspond to the same frequency $\omega_1 = \omega_2$. A linear combination of these solutions, $c_1\mathbf{x}^{(1)} + c_2\mathbf{x}^{(2)}$, where c_1 and c_2 are arbitrary coefficients, is also a solution with the same frequency. In other words, the set of solutions corresponding to a certain frequency forms a plane passing through the vectors $\mathbf{x}^{(1)}$ and $\mathbf{x}^{(2)}$, with any vector on this plane being orthogonal (in the mass or stiffness metric) to the normal vectors of oscillations corresponding to other frequencies. Among the vectors on this plane, we can choose a pair of independent vectors that satisfy the orthogonality relations (5); and we can do it in multiple ways.

The set of vectors that are mutually orthogonal (in the mass or stiffness metric) serves as a convenient basis for Q_α coordinates. These coordinates are referred to as *normal coordinates*, which we shall demonstrate by showing that these coordinates can reduce the Lagrangian to a form that corresponds to independent non-interacting oscillators.

In essence, a transition of the form (23.25) from the coordinates

$$\mathbf{x} = (x_1, \ldots, x_i, \ldots, x_s)$$

to normal coordinates

$$\mathbf{Q} = (Q_1, \ldots, Q_\alpha, \ldots, Q_s)$$

is a linear transformation

$$\mathbf{x} = \hat{U}\mathbf{Q}, \quad x_i = \sum_{\alpha} U_{i\alpha} Q_\alpha, \quad U_{i\alpha} \equiv A_i^{(\alpha)}.$$

This transformation simultaneously transforms the quadratic forms of the kinetic and potential energies to diagonal forms.

By substituting the transformation (23.25) into the Lagrangian (23.17) and utilizing the orthogonality properties (5), we observe that the Lagrangian can be rewritten as a sum of independent Lagrangians of the form (23.4) in new variables:

$$L = \sum_{\alpha=1}^{s} L_\alpha, \quad L_\alpha = \frac{1}{2}\left(M_\alpha \dot{Q}_\alpha^2 - K_\alpha Q_\alpha^2\right), \tag{24.6}$$

where

$$M_\alpha = \left(\mathbf{A}^{(\alpha)}, \hat{m}\,\mathbf{A}^{(\alpha)}\right) \quad \text{and} \quad K_\alpha = \left(\mathbf{A}^{(\alpha)}, \hat{k}\,\mathbf{A}^{(\alpha)}\right)$$

are the masses and stiffnesses, respectively. The corresponding Lagrangian equations take the form of one-dimensional equations (23.5):

$$M_\alpha \ddot{Q}_\alpha + K_\alpha Q_\alpha = 0. \tag{24.7}$$

Thus, each coordinate Q_α corresponds to a single oscillation with a distinct frequency ω_α, while each coordinate x_i may be composed as a linear combination of oscillations, each having distinct frequencies (see (23.26)).

It is worth noting that the eigenvalues of the roots of equation (23.22) are positive, owing to the positivity properties (23.18):

$$\omega_\alpha^2 = \frac{K_\alpha}{M_\alpha} = \frac{\left(\mathbf{A}^{(\alpha)}, \hat{k}\,\mathbf{A}^{(\alpha)}\right)}{\left(\mathbf{A}^{(\alpha)}, \hat{m}\,\mathbf{A}^{(\alpha)}\right)} \geq 0, \tag{24.8}$$

which implies that the eigenfrequencies ω_α are real. Furthermore, the eigenfrequencies defined by this formula are independent of the normalization of the vectors $\mathbf{A}^{(\alpha)}$.

24.3 *Oscillations of weakly coupled systems. Beats*

The motion of linear systems with frequency degeneration exhibits several interesting features that we will discuss using a simple example. Consider a mechanical system consisting of two weakly coupled subsystems, each capable of making small oscillations in the absence of coupling. At first glance, the motion of each subsystem in this case appears to occur almost independently, even in the presence of weak coupling. However, this is only true as long as the normal frequencies of the two subsystems do not coincide or are not close to each other. If these frequencies coincide or are close, then the influence of even weak coupling on the motion of the system turns out to be quite significant.

To illustrate this phenomenon, we consider an example of two mathematical pendulums of the same mass $m_1 = m_2 = m$ but different lengths l_1 and l_2 connected by a spring with a small elasticity coefficient k. When the pendulums hang vertically, the spring is

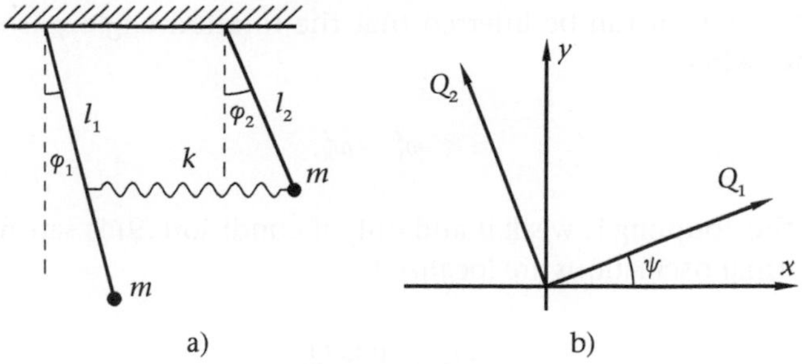

Figure 31 Weakly coupled systems: *a*) coupled pendulums, *b*) normal coordinates of weakly coupled pendulums ($x = l_1\varphi_1$, $y = l_2\varphi_2$)

horizontal and unstretched (Fig. 31a). If the angles of deviation of the pendulums from the vertical φ_1 and φ_2 are small, i.e. $|\varphi_{1,2}| \ll 1$, then the Lagrangian of the system can be expressed in the variables

$$x = l_1\varphi_1 \,, \quad y = l_2\varphi_2$$

as

$$L(x, y, \dot{x}, \dot{y}) = \frac{1}{2} m \left(\dot{x}^2 - \omega_x^2 x^2 + \dot{y}^2 - \omega_y^2 y^2 + 2\alpha xy\right) ,$$

where ω_x and ω_y are equal to

$$\omega_x = \sqrt{\frac{g}{l_1} + \frac{kl_2^2}{ml_1^2}} \,, \quad \omega_y = \sqrt{\frac{g}{l_2} + \frac{k}{m}} \,, \quad \alpha = \frac{kl_2}{ml_1} \,.$$

When $\alpha = 0$, we have two independent pendulums with so-called *partial frequencies* ω_x and ω_y.

Assume $l_1 \geq l_2$, such that $\omega_x \leq \omega_y$. At first glance, for $\alpha \ll \omega_x^2$, the coupling of these pendulums is weak, and the influence of this coupling on the motion of the system is insignificant. However, a direct calculation reveals that this is only true when $\alpha \ll \omega_y^2 - \omega_x^2$ is satisfied. Let us prove that.

It is evident that the transition to normal coordinates

$$Q_{1,2}(t) = a_{1,2} \cos(\omega_{1,2} t + \chi_{1,2})$$

corresponds to a rotation by the angle ψ in the xy-plane (see Fig. 31*b*):

$$x = Q_1 \cos\psi - Q_2 \sin\psi \,, \quad y = Q_1 \sin\psi + Q_2 \cos\psi \,,$$

$$\tan 2\psi = \frac{2\alpha}{\omega_y^2 - \omega_x^2} \,.$$

Additionally, the normal frequencies are given by

$$\omega_{1,2} = \frac{1}{\sqrt{2}} \left[\omega_x^2 + \omega_y^2 \mp \sqrt{\left(\omega_y^2 - \omega_x^2\right)^2 + 4\alpha^2}\right]^{1/2} ,$$

where $\omega_1 < \omega_x$ and $\omega_2 > \omega_y$.

From this expression, it can be inferred that the rotation angle ψ is small not when $\alpha \ll \omega_x^2$, but rather when

$$\alpha \ll \omega_y^2 - \omega_x^2. \tag{24.9}$$

In other words, the coupling is weak if and only if condition (9) is satisfied. In this case, $\psi \approx 0$ and the normal oscillations are localized:

$$x \approx Q_1\,, \quad y \approx Q_2\,,$$

while the normal frequencies are close to the partial ones:

$$\omega_1 \approx \omega_x, \quad \omega_2 \approx \omega_y.$$

However, if $\alpha \ll \omega_x^2$, but the partial frequencies are close to each other, and $\alpha \gg \omega_y^2 - \omega_x^2$, then $\psi \approx \pi/4$ and normal oscillations cease to be localized:

$$x \approx \frac{Q_1 - Q_2}{\sqrt{2}}\,, \quad y \approx \frac{Q_1 + Q_2}{\sqrt{2}}\,. \tag{24.10}$$

Under these conditions, *beats* may occur, which correspond to the significant transfer of the oscillation energy from one pendulum to the other. For example, let $l_1 = l_2 = l$ and consider the initial moment where only oscillations of the first pendulum are excited

$$x(0) = x_0, \; y(0) = \dot{x}(0) = \dot{y}(0) = 0\,. \tag{24.11}$$

Then the normal oscillations are

$$Q_1(t) = \frac{x_0}{\sqrt{2}} \cos \omega_1 t\,, \quad Q_2(t) = -\frac{x_0}{\sqrt{2}} \cos \omega_2 t$$

with

$$\omega_1 = \sqrt{\frac{g}{l}}\,, \quad \omega_2 = \sqrt{\frac{g}{l} + \frac{2k}{m}}\,.$$

The motions of the pendulums can then be expressed as:

$$\varphi_1(t) = \frac{x_0}{l} \cos \varepsilon t \cos \omega t\,, \quad \varphi_2(t) = \frac{x_0}{l} \sin \varepsilon t \sin \omega t\,, \tag{24.12}$$

where

$$\varepsilon = \frac{1}{2}(\omega_2 - \omega_1) \approx \frac{k}{2m}\sqrt{\frac{l}{g}} \ll \omega = \frac{1}{2}(\omega_2 + \omega_1) \approx \sqrt{\frac{g}{l}}\,.$$

It can be seen from Eq. (12) that solely the oscillations of the second pendulum will become exited after a time $\tau = \pi/(2\varepsilon)$. After a time of 2τ, the system will return to its initial state, and so on.

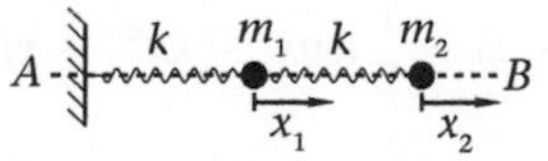

Figure 32 Particles moving along the straight line *AB*

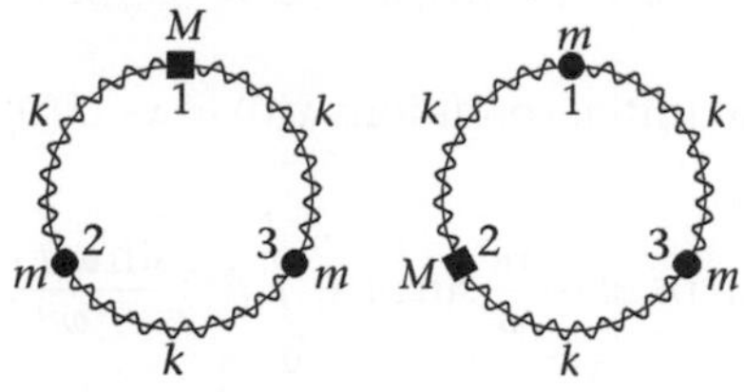

Figure 33 Particles on the ring

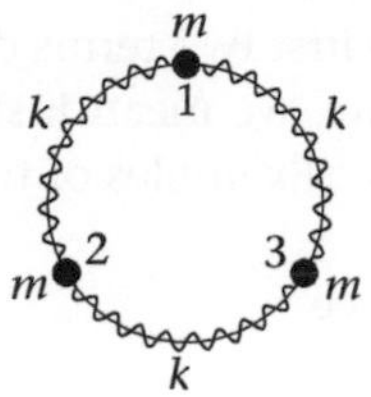

Figure 34 Symmetrical system

Problems

24.1. Find the free oscillations of the system shown in Fig. 32, in which the particles move only along the straight line *AB*. Consider the cases: a) $m_1 = m_2$; b) $m_1 \ll m_2$; c) $m_1 \gg m_2$. For the case *a*, find the normal coordinates and express the Lagrangian of the system in terms of them.

24.2. Find the normal oscillations of three particles (Fig. 33) on first and second rings.[2] The particles are connected by identical springs and able to move around the ring. Consider the transition $M \to m$.

24.3. Find the normal oscillations of three identical particles, connected by identical springs and able to move along the ring (Fig. 34). Find the free oscillations of this system, if, at the initial moment, the displacements of particles along the ring are $x_1(0) = -x_2(0) = a$, $x_3(0) = 0$, and the initial velocities are zero. Find the free oscillations under the same initial conditions for the system, obtained from the described one by changing the elasticity coefficient of the spring connecting particles 2 and 3 by a small amount δk.

§ 25 Forced oscillations. Resonances

Consider a one-dimensional oscillation system that is subjected to an external force $f(t)$, in addition to the elastic force $f_{elast} = -kx$. The additional potential energy can be expressed as $\Delta U(x, t) = -xf(t)$. The Lagrangian of such a system is given by

[2] In this and similar problems the ring is assumed to be smooth and fixed.

$$L(x, \dot{x}, t) = \frac{1}{2}(m\dot{x}^2 - kx^2) + xf(t)\,. \tag{25.1}$$

We write down the solution of the corresponding equation of motion

$$\ddot{x} + \omega^2 x = f(t)/m, \quad \omega = \sqrt{k/m} \tag{25.2}$$

for an arbitrary force $f(t)$ and initial conditions $x(0) = x_0$, $\dot{x}(0) = v_0$ in the form:

$$x(t) = x_0 \cos\omega t + \frac{v_0}{\omega}\sin\omega t + \int\limits_0^t f(\tau)\frac{\sin\omega(t-\tau)}{\omega m}d\tau\,. \tag{25.3}$$

It can be checked by a direct substitution. Here the last term corresponds to the *forced*, while the first two to the *free* oscillations.

In the presence of low friction, the first two terms disappear over time, while the latter changes only slightly. In what follows, we mean just such *steady-state oscillations* when speaking about the forced oscillations. Examples of transient processes can be found, for example, in [3], problem 5.11.

For a special case of a harmonic force

$$f(t) = f\cos(\gamma t + \varphi),$$

the forced oscillations have the form

$$x = b\cos(\gamma t + \varphi), \quad b = \frac{f}{(\omega^2 - \gamma^2)m}\,. \tag{25.4}$$

Dependence of the amplitude b on the frequency of the external force γ is shown in Fig. 35. For $\gamma < \omega$, oscillations are in phase with the acting force ($b/f > 0$), but when $\gamma > \omega$, they are in antiphase ($b/f < 0$). At $\gamma \gg \omega$, the oscillation amplitude is small ($b/f \ll 1/k$):

$$b \approx -\frac{f}{m\gamma^2}, \quad \gamma \gg \omega\,. \tag{25.4a}$$

At $\gamma \to \omega$ the amplitude $b \to \infty$, which is the case of *resonance*.

For cases involving many degrees of freedom, we can use normal coordinates by transforming the Lagrangian of the system from the original coordinates to normal

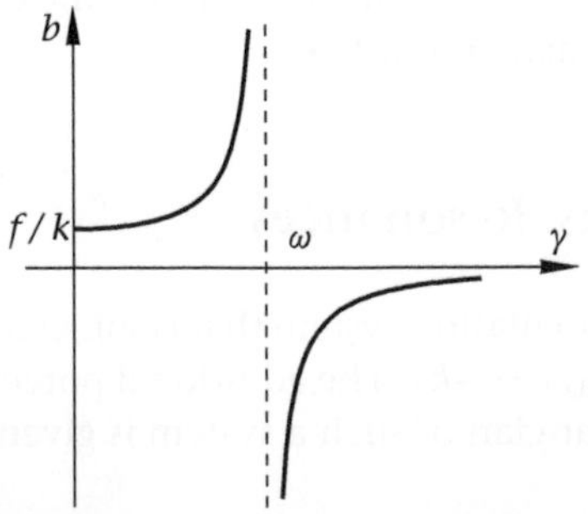

Figure 35 Dependence of the amplitude of forced oscillations b on the frequency of the driving force γ

coordinates. This reduces the original system to a set of one-dimensional oscillators that perform forced oscillations. The solution in this case is similar to the one-dimensional case.

Let the addition to the Lagrangian of free oscillations (23.17) have the form

$$\Delta L = \sum_i x_i F_i(t).$$

We introduce the force vector

$$\mathbf{F}(t) = \begin{pmatrix} F_1(t) \\ F_2(t) \\ \vdots \\ F_s(t) \end{pmatrix},$$

then

$$L(\mathbf{x}, \dot{\mathbf{x}}, t) = \frac{1}{2}\,(\dot{\mathbf{x}}), \hat{m}\,\dot{\mathbf{x}}) - \frac{1}{2}\,(\mathbf{x}, \hat{k}\,\mathbf{x}) + \mathbf{x}\mathbf{F}(t)\,; \tag{25.1a}$$

$$\hat{m}\,\ddot{\mathbf{x}} + \hat{k}\,\mathbf{x} = \mathbf{F}(t)\,. \tag{25.2a}$$

From this, we can transform the Lagrangian from the original coordinates to normal coordinates using Eq. (23.25). In this case

$$\mathbf{x}\mathbf{F}(t) = \sum_\alpha \mathbf{A}^{(\alpha)}\mathbf{F}(t) \cdot Q_\alpha$$

and the original Lagrangian transforms into a sum of Lagrangians of type (1):

$$L = \sum_{\alpha=1}^{s} L_\alpha\,, \quad L_\alpha = \frac{1}{2}\left(M_\alpha \dot{Q}_\alpha^2 - K_\alpha Q_\alpha^2\right) + Q_\alpha f_\alpha(t)\,, \tag{25.5}$$

where M_α and K_α are defined in (24.6) and the force

$$f_\alpha(t) = \mathbf{A}^{(\alpha)}\mathbf{F}(t)\,. \tag{25.6}$$

After this, the equations of motion for Q_α coincide with the one-dimensional Eq. (2):

$$\ddot{Q}_\alpha + \omega_\alpha^2 Q_\alpha = \frac{f_\alpha(t)}{M_\alpha}\,, \quad \omega_\alpha = \sqrt{\frac{K_\alpha}{M_\alpha}}\,. \tag{25.2b}$$

Let us consider the case of the harmonic force $\mathbf{F}(t) = \mathbf{F}\cos(\gamma t + \varphi)$; then

$$F_\alpha(t) = f_\alpha \cos(\gamma t + \varphi)\,, \quad f_\alpha = \mathbf{A}^{(\alpha)}\mathbf{F} \tag{25.7}$$

and forced oscillations of normal coordinates have the form

$$Q_\alpha = b_\alpha \cos(\gamma t + \varphi)\,, \quad b_\alpha = \frac{f_\alpha}{(\omega_\alpha^2 - \gamma^2)M_\alpha}\,. \tag{25.3a}$$

Switching from the normal coordinates to the original ones, we have

$$\mathbf{x} = \sum_{\alpha=1}^{s} \mathbf{A}^{(\alpha)} b_\alpha \cos(\gamma t + \varphi) =$$

$$= \sum_{\alpha=1}^{s} \mathbf{A}^{(\alpha)} \frac{(\mathbf{A}^{(\alpha)}, \mathbf{F})}{(\omega_\alpha^2 - \gamma^2)\left(\mathbf{A}^{(\alpha)}, \hat{m}\,\mathbf{A}^{(\alpha)}\right)} \cos(\gamma t + \varphi)\,. \tag{25.8}$$

Note that this result does not depend on the normalization of the vectors $\mathbf{A}^{(\alpha)}$.

The force $f_\alpha = \mathbf{A}^{(\alpha)}\mathbf{F}$ acting on the α-th normal oscillation is determined by the projection of the force vector $\mathbf{F}$ in the direction of this normal oscillation. Therefore, if $\mathbf{F}$ and $\mathbf{A}^{(\alpha)}$ are mutually orthogonal, that is $\mathbf{F}\mathbf{A}^{(\alpha)} = 0$, then the corresponding term $\mathbf{A}^{(\alpha)} b_\alpha \cos(\gamma t + \varphi)$ is absent in sum (8). In this case there is no resonance at $\gamma \to \omega_\alpha$.

However, if $\mathbf{F}\mathbf{A}^{(\alpha)} \neq 0$, then resonance arises at $\gamma \to \omega_\alpha$; moreover, near resonance, all terms in the sum (8), except for one, can be neglected and

$$\mathbf{x} \approx \mathbf{A}^{(\alpha)} b_\alpha \cos(\gamma t + \varphi).$$

Question. Let the vector $\mathbf{F}$ be parallel to some normal oscillation, for example,

$$\mathbf{F} = \text{const} \cdot \mathbf{A}^{(1)}.$$

It is clear that at $\gamma \to \omega_1$, resonance arises. Can such a force excite other normal oscillations (i.e. will the resonance occur for $\gamma \to \omega_2, \gamma \to \omega_3, \ldots, \gamma \to \omega_s$)?

For forced oscillations, the transition to normal coordinates is convenient since it reduces a multidimensional problem to a set of one-dimensional ones. In the particular case of a harmonic force, we can use a more direct approach. We are looking for a solution of Eq. (2a) in the form

$$\mathbf{x} = \mathbf{B} \cos(\gamma t + \varphi)\,.$$

To determine the amplitude $\mathbf{B}$, we obtain the equation

$$(-\gamma^2 \hat{m} + \hat{k})\mathbf{B} = \mathbf{F},$$

where

$$\mathbf{x} = (-\gamma^2 \hat{m} + \hat{k})^{-1}\, \mathbf{F} \cos(\gamma t + \varphi)\,. \tag{25.9}$$

Problem

25.1. Find the stable oscillations of the system described in problem 3.1*a* if the point A moves according to the law $a(t) = a \cos \gamma t$. At what frequency γ will the amplitude of the forced oscillations of the first particle vanish? Do the same for a chain of three identical particles, which is obtained if one attaches the third particle to the second particle using the same spring.

§ 26 Oscillations in the presence of a friction force

There is always friction present in macroscopic systems, which qualitatively changes both the normal oscillations and the resonant response to external harmonic forces. The nature of the friction force differs depending on whether sliding solids or motion of a rigid body in a liquid or gas is considered. In this section, we will only focus on the latter case.

Let us consider the effect of friction on the oscillations of a one-dimensional harmonic oscillator described by the coordinate x. To include friction force directly in Newton's second law, which is known from experience to be directed opposite and proportional to velocity at low absolute velocity values, we can represent the friction force as $f_{\text{fr}} = -\alpha\dot{x}$, where α is a positive constant that characterizes friction intensity. The corresponding equation of motion

$$m\ddot{x} = -\alpha\dot{x} - kx;$$

we divide by m and introduce the notation

$$\omega_0^2 = \frac{k}{m}, \quad \lambda = \frac{\alpha}{2m}.$$

Here ω_0 is the frequency of free oscillations in the absence of friction, and λ is *the damping coefficient*. The equation of free oscillations will be written in the form

$$\ddot{x} + 2\lambda\dot{x} + \omega_0^2 x = 0. \tag{26.1a}$$

Assuming

$$x = \text{Re}\left(e^{rt}\right)$$

we obtain a characteristic equation for r. Its roots are

$$r_{1,2} = -\lambda \pm \sqrt{\lambda^2 - \omega_0^2}.$$

The general solution is then a linear superposition of two solutions:

$$x = \text{Re}\left(C_1\, e^{r_1 t} + C_2\, e^{r_2 t}\right). \tag{26.2}$$

With strong friction ($\lambda > \omega_0$), both characteristic roots are negative, and the general solution x tends to zero without oscillations, termed as *aperiodic damping*.

For $\lambda = \omega_0$, the motion is again aperiodic:

$$x = (C_1 + C_2 t)\, e^{-\lambda t}.$$

For weaker friction ($\lambda < \omega_0$), the characteristic roots are complex conjugated,

$$r_{1,2} = -\lambda \pm i\omega, \quad \omega = \sqrt{\omega_0^2 - \lambda^2},$$

and the solution is given by:

$$x = ae^{-\lambda t}\cos(\omega t + \varphi), \tag{26.3}$$

where a and φ are real constants. Such a motion represents *damped oscillations*.

In the important limiting case $\lambda \ll \omega_0$, the amplitude of oscillations almost does not change during the time $T = 2\pi/\omega$. The squares of coordinate and velocity averaged over this time can be calculated by counting the factor $e^{-\lambda t}$ to be constant. These average squares are proportional to $e^{-2\lambda t}$ and the average energy of oscillations decreases according to the same law:

$$E(t) = E_0\, e^{-2\lambda t}\,.$$

Now, let us consider the effect of friction on the motion of the oscillator under the action of an external harmonic force $f\cos(\gamma t + \varphi)$. The equation of motion has the form

$$\ddot{x} + 2\lambda\dot{x} + \omega_0^2 x = \frac{f}{m}\cos(\gamma t + \varphi)\,. \tag{26.1b}$$

The solution of this inhomogeneous linear equation is the sum of two terms: the free motion described above and forced (stable) oscillations. The latter solution is convenient to search in complex form. For this we write

$$\cos(\gamma t + \varphi) = \mathrm{Re}\left[e^{i(\gamma t+\varphi)}\right].$$

By looking for a solution in the form:

$$x = \mathrm{Re}\left[B e^{i(\gamma t+\varphi)}\right],$$

we find the complex amplitude

$$B = b e^{i\delta} = \frac{f}{m(\omega_0^2 - \gamma^2 + 2i\lambda\gamma)}\,.$$

The imaginary component of B is consistently negative, whereas the real component changes sign from positive to negative as γ increases from zero to infinity, occurring at $\gamma = \omega_0$. This alteration results in the argument δ taking values in the range $0 > \delta > -\pi$.

For the real amplitude b and argument δ we have

$$b = \frac{f}{m\sqrt{(\omega_0^2 - \gamma^2)^2 + 4\lambda^2\gamma^2}}, \qquad \tan\delta = \frac{2\lambda\gamma}{\gamma^2 - \omega_0^2}\,. \tag{26.4}$$

From these equations, the amplitude of forced oscillations b is observed to have its maximum value (for a given force amplitude f) at frequency γ_{m}, where

$$\gamma_{\mathrm{m}} = \sqrt{\omega_0^2 - 2\lambda^2} \tag{26.5}$$

and is equal to

$$b_{\max} = b(\gamma_{\mathrm{m}}) = \frac{f}{2m\lambda\sqrt{\omega_0^2 - \lambda^2}}\,. \tag{26.6}$$

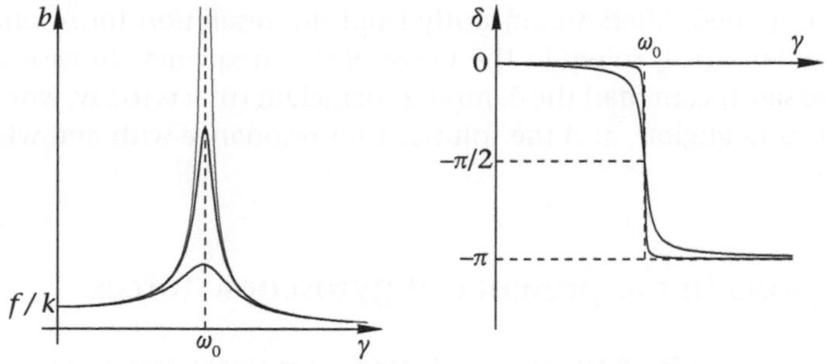

Figure 36 Dependence of the amplitude of forced oscillations b and phase δ on the driving force frequency γ for different values of the damping coefficient λ

Figure 36 displays the graphical relation between the functions b and δ concerning the frequency of the driving force γ for various damping coefficients λ. For small values of λ, the oscillations are similar to that of the undamped system, as demonstrated in Fig. 35. Specifically, at $\gamma < \omega_0$, δ approaches zero, indicating that the forced oscillations follow the acting force in-phase, whereas for $\gamma > \omega_0$, δ approximates $(-\pi)$ and the forced oscillations occur almost in antiphase with the applied force. The transition between these two phases occurs within a very narrow frequency range near ω_0.

Free oscillations damp, and only forced oscillations remain with time

$$x = b\cos(\gamma t + \varphi + \delta)\,.$$

Since the phase of δ is negative, the cosine argument of this solution has the value that the cosine argument of the driving force had at an earlier moment of time; they say that forced oscillation lags behind the driving force.

In the case of low damping, $\lambda \ll \omega_0$, we examine the resonance vicinity. Denoting the *detuning frequency* as

$$\epsilon = \gamma - \omega_0$$

and using appropriate approximations in Eq. (4),

$$\gamma^2 - \omega_0^2 = (\gamma + \omega_0)(\gamma - \omega_0) \approx 2\omega_0\epsilon, \quad 2\lambda\gamma \approx 2\lambda\omega_0\,, \tag{26.7}$$

we get

$$b = \frac{f}{2m\omega_0\sqrt{\epsilon^2 + \lambda^2}}\,, \quad \tan\delta = \frac{\lambda}{\epsilon}\,. \tag{26.8}$$

The maximum amplitude of forced oscillations is inversely proportional to λ:

$$b_{\max} = \frac{f}{2m\omega_0\lambda}\,. \tag{26.9}$$

At $\epsilon = \pm\lambda$, the oscillation amplitude squared is half the maximum value. The phase change δ primarily occurs within the same frequency range. Therefore, the resonant region's characteristic width is λ.

The solution obtained differs significantly from the resolution for resonance without friction given by Eq. (25.4) solely in the range of this frequency. In case the frequency detuning is more significant than the damping coefficient (that is to say, when $|\epsilon| \gg \lambda$), the effect of friction is negligible, and the solutions for resonance with and without friction are almost equal.

§ 27 Oscillations in the presence of gyroscopic forces

In § 23 and § 24 we considered the free oscillations of systems that move under the influence of potential forces. In this case, linear oscillations occur when the systems deviate slightly from the equilibrium position, which is typically near the minimum of the potential energy function $U(\mathbf{x})$. A single normal oscillation (23.24) takes place along a straight line, such that $\mathbf{x} = \mathbf{A}\cos(\omega t + \varphi)$.

In this section we discuss the linear oscillations of systems that experience not only potential forces, but also non-potential gyroscopic forces $\mathbf{F}_g$. *Gyroscopic forces are linear in particle velocity and orthogonal to this velocity.*

27.1 *Gyroscopic forces*

We start from the Lagrangian for a particle of the form

$$L(\mathbf{r}, \mathbf{v}) = L_0 + L_g\,, \quad L_0 = \frac{1}{2} m\mathbf{v}^2 - U(\mathbf{r})\,, \quad L_g = \mathbf{v} \cdot \mathbf{C}(\mathbf{r})\,, \tag{27.1}$$

where $\mathbf{C}(\mathbf{r})$ is a known function of the coordinates $\mathbf{r}$. The term L_0 refers to the particle's motion in the potential field $U(\mathbf{r})$, while the term L_g gives rise to the gyroscopic force $\mathbf{F}_g$. Specifically, the Lagrangian equations

$$\frac{d}{dt}\frac{\partial L}{\partial \mathbf{v}} - \frac{\partial L}{\partial \mathbf{r}} = 0$$

can be expressed as

$$m\ddot{\mathbf{r}} = -\nabla U(\mathbf{r}) + \mathbf{F}_g\,, \quad \mathbf{F}_g = [\mathbf{v},\, [\nabla,\, \mathbf{C}(\mathbf{r})]]\,. \tag{27.2}$$

Given that $\mathbf{v} \cdot \mathbf{F}_g = 0$, the work done by the gyroscopic force equals zero. Consequently, the energy

$$E = \frac{1}{2} m\mathbf{v}^2 + U(\mathbf{r})$$

is conserved.

To provide examples of gyroscopic forces, let us consider the following cases:

(i) when a particle with a charge e moves in a constant magnetic field $\mathbf{B}(\mathbf{r})$ (see § 12), the function

$$\mathbf{C}(\mathbf{r}) = \frac{e}{c}\mathbf{A}(\mathbf{r})\,,$$

where $\mathbf{A}(\mathbf{r})$ is the vector potential. The gyroscopic force coincides with the Lorentz force:

$$\mathbf{F}_g = \frac{e}{c}[\mathbf{v}, \mathbf{B}(\mathbf{r})]\,.$$

(ii) when a particle moves in a reference frame rotating with a constant angular velocity $\boldsymbol{\Omega}$ (see § 20.2), the function

$$\mathbf{C}(\mathbf{r}) = m\,[\boldsymbol{\Omega}, \mathbf{r}]\,.$$

The gyroscopic force is the same as the Coriolis force:

$$\mathbf{F}_g = 2m[\mathbf{v}, \boldsymbol{\Omega}]\,.$$

Linear oscillations in the presence of gyroscopic forces exhibit some interesting features. For instance, these oscillations can occur not only near the minimum of the potential energy function $U(\mathbf{x})$, but also near its maximum. Moreover, the trajectory of oscillations for a given frequency need not resemble a straight line (23.24). The following section provides some illustrative examples.

27.2 *Small oscillations of a charged particle in a magnetic field*

Let the potential energy near the extremum be a diagonal quadratic form of coordinates

$$U(x,y,z) = \frac{1}{2}k_x x^2 + \frac{1}{2}k_y y^2 + \frac{1}{2}k_z z^2\,, \tag{27.3}$$

where the coefficient k_z is positive,

$$k_z > 0\,,$$

and the coefficients $k_{x,y}$ can be either positive or negative. If k_x and k_y are positive, the potential field is referred to as an *oscillatory* field, where the minimum potential energy is found near the coordinate origin. Additionally, the equipotential surface is similar to a well in this region at $z = 0$. On the other hand, if k_x and k_y are both negative, the potential field is referred to as an *anti-oscillatory* field, where the potential energy has a maximum at the point $x = y = 0$. In this case, the equipotential surface is similar to a hill. Finally, if k_x and k_y have opposite signs, the equipotential surface near the origin has the form of a saddle.

Furthermore, a constant uniform magnetic field is assumed to be directed along the z-axis:

$$\mathbf{B} = (0,\, 0,\, B)$$

and the vector potential is chosen in the form

$$\mathbf{A}(\mathbf{r}) = \frac{1}{2}\,[\mathbf{B}, \mathbf{r}] = \frac{1}{2}\,B\,(-y,\, x,\, 0)\,. \tag{27.4}$$

The Lagrangian of this system is a quadratic form of Cartesian coordinates and velocities

$$L(\mathbf{r}, \mathbf{v}) = \frac{1}{2}\,m\left(\dot{x}^2 + \dot{y}^2 + \dot{z}^2\right) - \frac{1}{2}\left(k_x x^2 + k_y y^2 + k_z z^2\right) + \frac{eB}{2c}\,(x\dot{y} - y\dot{x})\,. \tag{27.5}$$

It is impossible to transform this Lagrangian function into a diagonal form using only linear transformation of the coordinates. The transition to normal coordinates is connected to the canonical transformation – see problem 11.9 from [3].

The magnetic field does not affect the motion along the z-axis, so the movement corresponds to the free oscillations:

$$z = a_z \cos(\omega_z t + \varphi_z), \quad \omega_z = \sqrt{k_z/m}. \tag{27.6}$$

Let us consider in more detail the motion of a particle in the xy-plane. The potential energy $U(x, y, 0)$ has an extremum at the point $x = y = 0$. Denoting

$$\omega_{\mathrm{B}} = \frac{eB}{mc}, \quad k_B = m\omega_{\mathrm{B}}^2 \tag{27.7}$$

we represent the equations of particle motion in the form:

$$\begin{aligned} m\ddot{x} + k_x x - m\omega_{\mathrm{B}}\dot{y} &= 0, \\ m\ddot{y} + k_y y + m\omega_{\mathrm{B}}\dot{x} &= 0. \end{aligned}$$

The solutions we seek for these equations appear in the form of harmonic oscillations

$$x = \mathrm{Re}(Ae^{i\omega t}), \quad y = \mathrm{Re}(Ce^{i\omega t}),$$

where the complex amplitude $A = ae^{i\varphi}$, while the quantities a and φ are real. These assumptions reduce the equations of motion to a system of homogeneous equations for the amplitudes A and C:

$$\begin{aligned} (k_x - m\omega^2)A - im\omega_{\mathrm{B}}\omega C &= 0, \\ im\omega_{\mathrm{B}}\omega A + (k_y - m\omega^2)C &= 0. \end{aligned}$$

For the system to have non-trivial solutions, the determinant of the system should be zero:

$$m^2\omega^4 - m\omega^2(k_B + k_x + k_y) + k_x k_y = 0.$$

The roots of this equation are

$$\omega_{1,2}^2 = \frac{1}{2m}\left[k_B + k_x + k_y \pm \sqrt{(k_B + k_x + k_y)^2 - 4k_x k_y}\right] \tag{27.8}$$

and they satisfy the conditions according to the Vieta theorem:

$$\omega_1^2\,\omega_2^2 = \frac{k_x\,k_y}{m^2}, \quad \omega_1^2 + \omega_2^2 = \frac{k_B + k_x + k_y}{m}. \tag{27.9}$$

Furthermore, these roots are positive under one of two conditions:

1) The field is oscillatory, in which case

$$k_x > 0, \; k_y > 0. \tag{27.10}$$

In this scenario, the potential energy has a minimum at $x = y = 0$. This case is discussed in § 27.3;

2) The field is anti-oscillatory with

$$k_x < 0, \; k_y < 0, \tag{27.11}$$

and the magnetic field is sufficiently large such that

$$k_B > |k_x| + |k_y| + 2\sqrt{k_x k_y} = \left(\sqrt{|k_x|} + \sqrt{|k_y|}\right)^2 . \tag{27.12}$$

Here, the potential energy has a maximum at $x = y = 0$, which is discussed in § 27.4.

Under either condition, the solutions take the form of oscillations near the origin of coordinates:

$$x = a_\alpha \cos(\omega_\alpha t + \varphi_\alpha) , \quad y = b_\alpha \sin(\omega_\alpha t + \varphi_\alpha) , \tag{27.13}$$

where $\alpha = 1, 2$ and enumerates two different eigen oscillations. The coefficients a_α and b_α are given by:

$$b_\alpha = \frac{m\omega_B \omega_\alpha}{k_y - m\omega_\alpha^2} a_\alpha = \frac{k_x - m\omega_\alpha^2}{m\omega_B \omega_\alpha} a_\alpha . \tag{27.14}$$

Free oscillations of systems moving under the action of only potential forces correspond to small deviations from equilibrium positions near the minimum of potential energy. In this case, a single normal oscillation occurs along a straight line. In the considered case, the motion of a particle in the xy-plane is a superposition of the found oscillations (13). These oscillations can be described as normal, thus generalizing the concept of normal oscillation. The motions in the x- and y-axes occur with the same frequency, but with a phase shift of $\pm\pi/2$, i.e. not along a straight line, but along an ellipse.

If conditions (10) or (11)–(12) are not satisfied, at least one of the roots $\omega_{1,2}^2$ becomes non-positive, corresponding to the particle departing from the origin. Specifically, for $k_x < 0$ and $k_y > 0$, the equipotential surface near the origin (with $z = 0$) takes the form of a saddle, and the typical particle trajectory in this case is shown in Fig. 37.

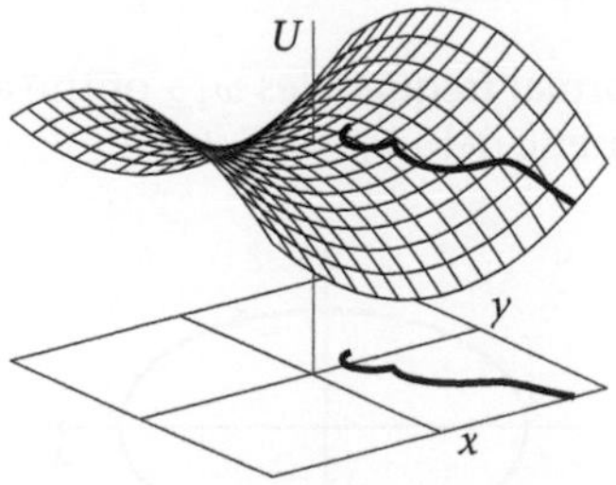

Figure 37 The particle leaves the saddle point of potential energy moving roughly along the equipotential line

27.3 *Oscillator in a uniform magnetic field*

In this scenario, we have $k_x > 0$, $k_y > 0$ and potential energy that corresponds to an attractive field. We can denote the following:

$$\omega_x = \sqrt{\frac{k_x}{m}}, \quad \omega_y = \sqrt{\frac{k_y}{m}}$$

and assume for definiteness that $\omega_x > \omega_y$ and $\omega_{\mathrm{B}} > 0$. The normal oscillation frequencies (8) can then be expressed as:

$$\omega_{1,2} = \frac{1}{2}\left(\sqrt{\omega_B^2 + \omega_+^2} \pm \sqrt{\omega_B^2 + \omega_-^2}\right), \quad \omega_\pm = \omega_x \pm \omega_y .$$

The dependence of the normal frequencies $\omega_{1,2}$ on the magnetic field is shown in Fig. 38. The limiting cases are as follows:

$$\omega_1 \to \omega_x, \quad \omega_2 \to \omega_y \quad \text{at } \omega_{\mathrm{B}} \ll \omega_x - \omega_y,$$

$$\omega_1 \to \omega_{\mathrm{B}}, \quad \omega_2 \to \frac{\omega_x \omega_y}{\omega_{\mathrm{B}}} \quad \text{at } \omega_{\mathrm{B}} \gg \omega_x .$$

From equation (14), we can obtain the amplitudes a_1 and b_1 of the normal oscillation with the high frequency ω_1:

$$\frac{b_1}{a_1} = \frac{\omega_x^2 - \omega_1^2}{\omega_{\mathrm{B}}\omega_1} < 0, \quad \left(\frac{b_1}{a_1}\right)^2 = \frac{\omega_1^2 - \omega_x^2}{\omega_1^2 - \omega_y^2} < 1 .$$

These relations indicate that the normal oscillation follows a clockwise motion along an ellipse with a major axis directed along the *x*-axis, as shown in Fig. 39. Similarly, for the normal oscillation with low frequency ω_2, we have:

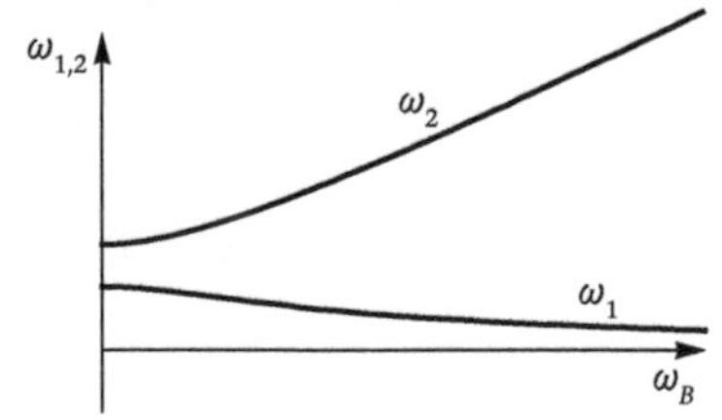

Figure 38 Dependence of normal frequencies $\omega_{1,2}$ of the anisotropic charged oscillator on the magnitude of the magnetic field $\omega_{\mathrm{B}} = eB/(mc)$

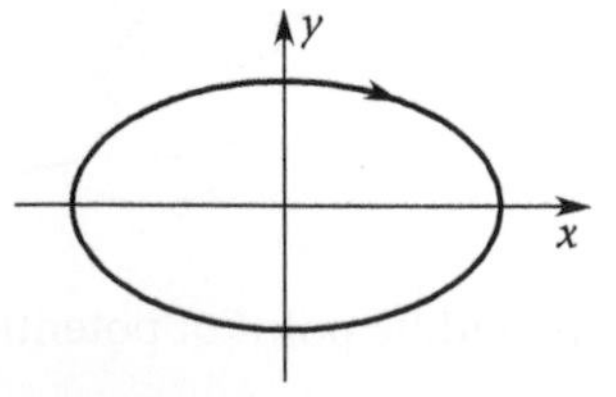

Figure 39 High frequency normal oscillation ω_1 for oscillator in a magnetic field

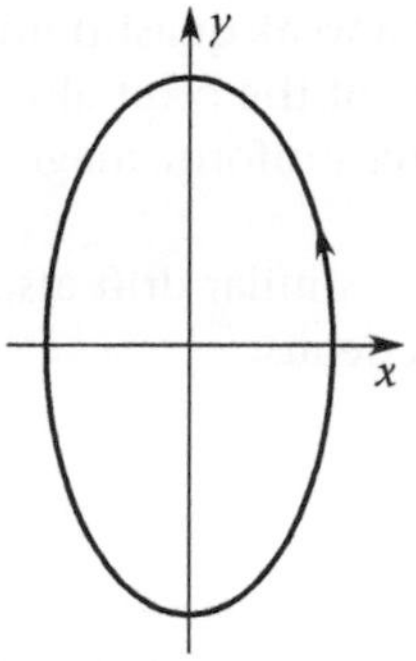

Figure 40 Low frequency normal oscillation ω_2 for oscillator in a magnetic field

$$\frac{b_2}{a_2} = \frac{\omega_x^2 - \omega_2^2}{\omega_B \omega_2} > 0, \quad \left(\frac{b_2}{a_2}\right)^2 = \frac{\omega_x^2 - \omega_2^2}{\omega_y^2 - \omega_2^2} > 1,$$

which indicates a counter clockwise motion along an ellipse with the major axis directed along the y-axis, as shown in Fig. 40.

We will now examine three limiting cases in greater detail.

1) Weak magnetic field

When

$$\omega_B \ll \omega_x - \omega_y \neq 0,$$

Larmor's theorem, which applies only to symmetry about the z-axis fields, is not applicable. As a result, the ellipses of normal oscillations are considerably elongated, and the normal oscillation frequencies are

$$\omega_{1,2} \approx \omega_{x,y} \pm \frac{\omega_B^2 \omega_{x,y}}{2(\omega_x^2 - \omega_y^2)}$$

which are close to $\omega_{x,y}$.

2) Strong magnetic field

In the case of

$$\omega_B \gg \omega_{x,y},$$

a normal oscillation with frequency $\omega_1 \approx \omega_B$ occurs along a circle. On the other hand, a normal oscillation with frequency $\omega_2 \approx \omega_x \omega_y / \omega_B$ occurs along an ellipse whose ratio of semi-axes $b_2/a_2 = \omega_x/\omega_y$ is the same as that of the equipotential surfaces. The equipotential surface $U(x, y, 0) = \text{const} = \frac{m}{2} N^2$ can be represented as an ellipse

$$\frac{x^2}{(N/\omega_x)^2} + \frac{y^2}{(N/\omega_y)^2} = 1$$

with semi-axes N/ω_x and N/ω_y along the x- and y-axes. Hence, the motion is circular, while the centre of the circle moves or drifts slowly along an ellipse.

It is known that the appearance of a weak quasi-homogeneous potential field $U(\mathbf{r})$ leads to a slow displacement of the centre of the orbit along the level line of field $U(\mathbf{r})$ when a charged particle moves in a strong uniform magnetic field in a plane perpendicular to it.

Note that, in the considered case, a similar drift also occurs even in the case when the oscillatory field is not quasi-homogeneous.

3) Isotropic oscillator

In an isotropic oscillator, where

$$\omega_x = \omega_y \equiv \omega,$$

the normal oscillations in the xy-plane correspond to circular motions with opposite directions and frequencies given by

$$\omega_{1,2} = \tilde{\omega} \pm \frac{1}{2}\omega_{\mathrm{B}}\,, \quad \tilde{\omega} = \sqrt{\omega^2 + \frac{1}{4}\omega_{\mathrm{B}}^2}.$$

In a rotating system with a frequency of $(-\frac{1}{2}\omega_{\mathrm{B}})$, both of these motions have frequencies of $\tilde{\omega}$, which represents the normal oscillation frequency of the isotropic oscillator. Indeed, the sum and difference of such oscillations with equal amplitudes

$$\begin{pmatrix} \cos\tilde{\omega}t \\ -\sin\tilde{\omega}t \end{pmatrix} \pm \begin{pmatrix} \cos\tilde{\omega}t \\ \sin\tilde{\omega}t \end{pmatrix}$$

are linear oscillations along the x- or y-axes.

In the limit where the magnetic field is small, $\omega_{\mathrm{B}} \ll \omega$, the effect of the magnetic field is only to cause rotation ('precession') around the z-axis with a frequency of $(-\frac{1}{2}\omega_{\mathrm{B}})$ (see Larmor's theorem in § 20.3). However, if $\omega_{\mathrm{B}} \gtrsim \omega$, the use of the rotating system loses visibility.

27.4 *Anti-oscillator in a uniform magnetic field*

In the case where $k_x < 0$ and $k_y < 0$ for the anti-oscillator, we can use the notation

$$\gamma_x = \sqrt{\frac{|k_x|}{m}}, \quad \gamma_y = \sqrt{\frac{|k_y|}{m}}$$

and assume for definiteness that $\gamma_x > \gamma_y$. Thus, instead of the field of attraction, we consider now the field of repulsion, but at a sufficiently large magnetic field (cf. (12))

$$\omega_{\mathrm{B}} > \gamma_x + \gamma_y\,. \tag{27.15}$$

The frequencies of normal oscillations (8) in this case can be presented in the form

$$\omega_{1,2} = \frac{1}{2}\left(\sqrt{\omega_B^2 - \gamma_-^2} \pm \sqrt{\omega_B^2 - \gamma_+^2}\right), \quad \gamma_\pm = \gamma_x \pm \gamma_y\,. \tag{27.16}$$

A typical particle trajectory in such a field is shown in Fig. 41, where stability near the maximum potential energy is provided by a sufficiently large Lorentz force.

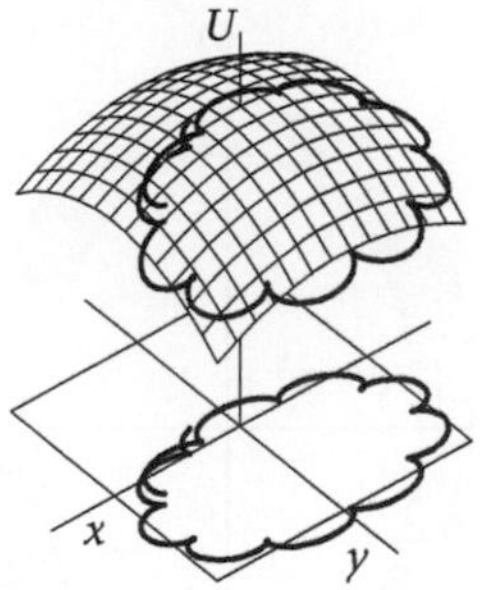

Figure 41 The Lorentz force prevents a particle from falling off the potential hill

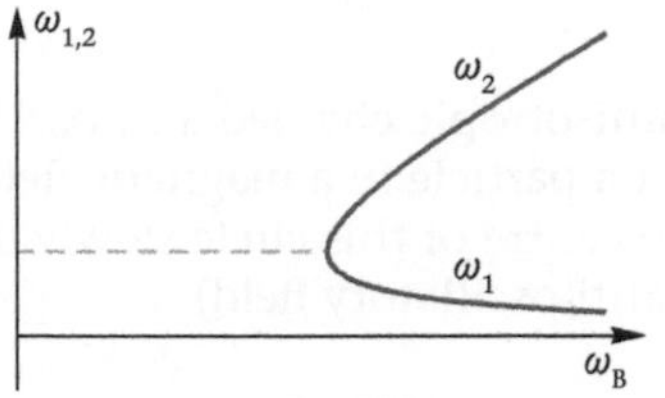

Figure 42 Dependence of normal frequencies $\omega_{1,2}$ of the charged anti-oscillator on the magnitude of the magnetic fields $\omega_{\mathrm{B}} = eB/(mc)$

The dependence of the normal frequencies $\omega_{1,2}$ on the magnitude of the magnetic field is shown in Fig. 42, the limiting cases being

$$\omega_{1,2} \;\to\; \sqrt{\gamma_x\gamma_y}\,, \qquad \text{when } \omega_{\mathrm{B}} \to \gamma_x + \gamma_y\,,$$

$$\omega_1 \;\to\; \omega_{\mathrm{B}}\,,\; \omega_2 \to \frac{\gamma_x\gamma_y}{\omega_{\mathrm{B}}} \quad \text{when } \omega_{\mathrm{B}} \gg \gamma_x\,.$$

The following relations take place for the amplitudes of the α-th normal oscillation (see Eq. (14)):

$$\frac{b_\alpha}{a_\alpha} = -\frac{\omega_\alpha^2 + \gamma_x^2}{\omega_{\mathrm{B}}\omega_\alpha} < 0\,, \quad \left(\frac{b_\alpha}{a_\alpha}\right)^2 = \frac{\omega_\alpha^2 + \gamma_x^2}{\omega_\alpha^2 + \gamma_y^2} > 1\,, \quad \alpha = 1, 2\,. \tag{27.17}$$

They show that both normal oscillations are clockwise motions along ellipses with a major axis, directed along the y-axis. For the case of a strong magnetic field the trajectory is shown in Fig. 43.

27.5 *Penning trap*

The preceding discussion illuminates the principles underlying the operation of the Penning trap,[3] which is depicted in Fig. 44. This device is capable of confining a single ion or electron for an extended period of time, enabling ultra-precise measurements of these particles' magnetic moments. For their pioneering research in this area, Hans Dehmelt and Wolfgang Paul received the Nobel Prize in Physics in 1989.

[3] You can read more about this trap, for example, in book [12], from where specific numbers are taken.

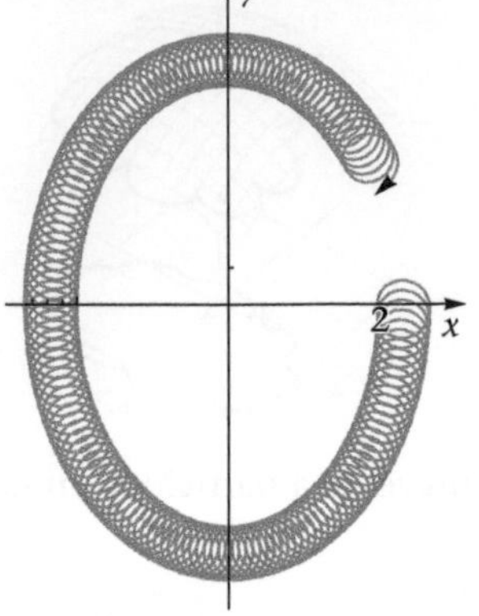

Figure 43 Trajectory of an anisotropic charged anti-oscillator in a strong magnetic field. The case is shown when a particle in a magnetic field rotates clockwise around a circle of small radius, and the centre of this circle slowly drifts, also clockwise, along the ellipse (equipotential of the anti-oscillatory field)

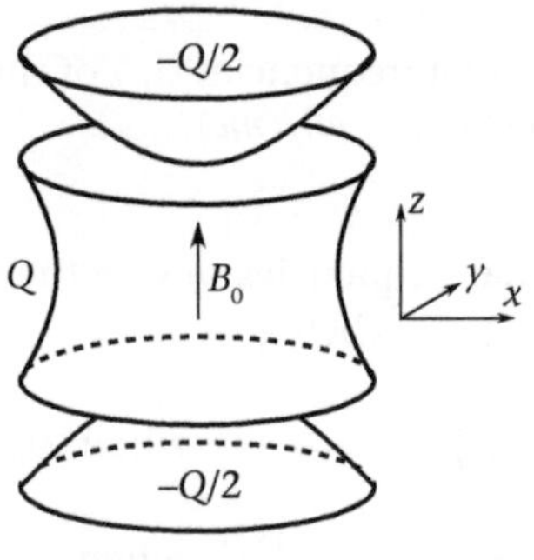

Figure 44 Scheme of a Penning trap

To illustrate the mechanism of charged particle confinement (specifically, electron confinement) in a Penning trap, we consider the following schematic. The trap employs a combination of electric and magnetic fields. A strong magnetic field $\mathbf{B}_0$ (5 tesla) is directed along the trap's central axis (z-axis), while electrodes create an electric field, forming a quadrupole with potential energy characterized by parameters

$$k_z = -2k_x = -2k_y > 0\,,$$

as given by Eq. (3).

Electrons within the trap exhibit simple harmonic motion along the magnetic field's axis (i.e. the z-axis), with a corresponding oscillation frequency defined by equation (6). Specifically,

$$\frac{\omega_z}{2\pi} = \frac{\sqrt{k_z/m_e}}{2\pi} = 60\ \text{MHz}.$$

In the xy-plane, the trap forms an anti-oscillator in a magnetic field, as discussed in the preceding section. Repulsive forces correspond to the quadrupole electric field. The electron's motion is stabilized in this plane by the Lorentz force generated by the powerful magnetic field with parameters

$$\gamma_x = \gamma_y \equiv \gamma = \sqrt{\frac{k_z}{2m_e}}, \quad \omega_B = -\frac{|e|B_0}{m_e c}$$

provided $|\omega_B| \gg \gamma$. Typical values for the normal frequencies in this plane are

$$\frac{\omega_1}{2\pi} \approx \frac{|\omega_B|}{2\pi} = 140 \text{ GHz}, \quad \frac{\omega_2}{2\pi} \approx \frac{\omega_z^2/(2|\omega_B|)}{2\pi} \approx 13 \text{ kHz}.$$

Thus, stability of motion along the trap's axis is provided by an electric field, whereas magnetic field-induced Lorentz forces stabilize movement in the xy-plane. The electron follows a fast, counterclockwise circular trajectory within a circle of small radius, while the centre of the trapped electron's motion drifts slowly in a counterclockwise manner within a circle of much larger radius. This design permits the confinement of an electron in the trap for many months! Fig. 43 closely resembles the trajectory that the electron follows in the xy-plane, albeit with some modifications owing to the electron's negative charge.

27.6 *Particle inside a smooth rotating paraboloid in the field of gravity*

We consider a particle inside a smooth paraboloid

$$z = \frac{x^2}{2a} + \frac{y^2}{2b},$$

rotating around the z-axis with a constant angular velocity Ω, in the field of gravity $\mathbf{g} = (0, 0, -g)$. We want to find the value of Ω at which the lower position becomes unstable for a particle inside the paraboloid.

Let $\mathbf{r}$ and $\mathbf{v}$ be the radius vector and the velocity of the particle in rotating coordinate system, respectively. In this coordinate system, the Lagrangian function is given by (see § 20.2)

$$\begin{aligned} L(\mathbf{r}, \mathbf{v}) =& \frac{1}{2} m \left(\mathbf{v} + [\boldsymbol{\Omega}, \mathbf{r}]\right)^2 + m\mathbf{g}\mathbf{r} = \\ =& \frac{1}{2} m \left[(\dot{x} - \Omega y)^2 + (\dot{y} + \Omega x)^2 + \dot{z}^2 - g\left(\frac{x^2}{a} + \frac{y^2}{b}\right)\right]. \end{aligned}$$

For small oscillations near the origin, we can neglect the term

$$\frac{1}{2} m \dot{z}^2 = \frac{1}{2} m \left(\frac{x}{a}\dot{x} + \frac{y}{b}\dot{y}\right)^2 \ll \frac{1}{2} m(\dot{x}^2 + \dot{y}^2)$$

and consider only the motion in the xy-plane. In this case, the Lagrangian of this problem differs from the Lagrangian (5) by notation only:

$$\frac{k_x}{m} = \frac{g}{a} - \Omega^2, \quad \frac{k_y}{m} = \frac{g}{b} - \Omega^2, \quad \omega_B = 2\Omega.$$

Now it is easy to verify that the motion near the origin will be stable at

$$\left(\frac{g}{a} - \Omega^2\right)\left(\frac{g}{b} - \Omega^2\right) > 0,$$

while under the condition

$$\left(\frac{g}{a} - \Omega^2\right)\left(\frac{g}{b} - \Omega^2\right) < 0$$

the particle moves away from the origin. Assuming for definiteness $a > b$, we get the region of instability

$$\frac{g}{a} < \Omega^2 < \frac{g}{b}.$$

Note that even for $\Omega^2 > g/b$, the motion is stable although the potential energy in a rotating reference frame

$$U = -\frac{m}{2}\left(\Omega^2 - \frac{g}{a}\right)x^2 - \frac{m}{2}\left(\Omega^2 - \frac{g}{b}\right)y^2$$

represents not a potential well, but a potential hill. Stability in this case is provided by the action of the Coriolis force.

27.7 *Lagrange points in the solar system*

This is an interesting example of the manifestation of gyroscopic forces in celestial mechanics: the motion of groups of asteroids under the influence of the Sun and Jupiter near the so-called Lagrange points.[4] For simplicity, we will assume that Jupiter and the Sun move in a circular motion around their centre of mass, and the influence of other planets is negligible (Fig. 45). We use a frame of reference that rotates with the same angular velocity as Jupiter and the Sun and has its centre at the centre of mass of the Sun–Jupiter system. Each asteroid in a group is affected by potential forces of attraction to the Sun and Jupiter and repulsive centrifugal force of inertia. Lagrange points are those points where the sum of these forces is zero. Two such points move along the Jupiter orbit at 60° ahead

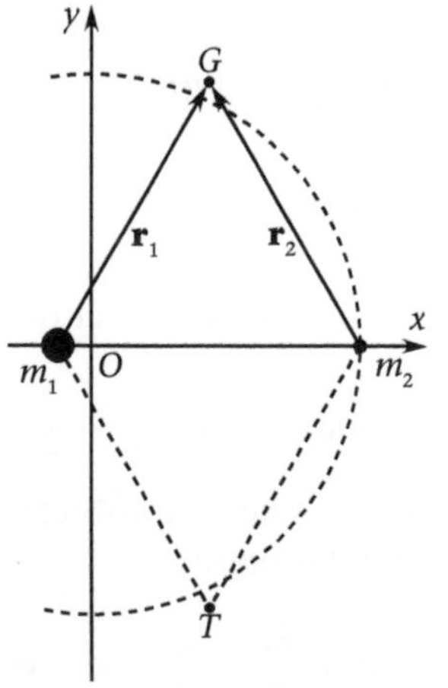

Figure 45 Asteroid of small mass m in the field of the Sun (mass m_1) and Jupiter (mass m_2)

[4] More details on this subject can be read, for example, in [2], section 3.12.

and behind it (see points *G* and *T* on Fig. 45). In other words, the Sun, Jupiter, and either of the Lagrange points form a regular triangle.

As a function of the asteroid coordinates defining position in the plane of the orbit, potential energy has its maximum near these Lagrange points (see problem 9.29 from [3]). However, the motion of the asteroid near these points is stable due to the influence of the Coriolis force.

This phenomenon can be observed in the clusters of asteroids that exist near these Lagrange points known as 'Greeks' and 'Trojans'.

§ 28 Oscillations of symmetric systems

Many mechanical and electrical systems, as well as the crystal lattice of solid bodies and molecules, exhibit certain symmetry properties. In this section, we present examples of such symmetrical molecules. Figure 46 shows the linear molecule CO_2 which does not change when rotated over 180° around the y-axis. The planar molecule C_2H_4 (Fig. 47) does not change when rotated at 180° around the x-axis or around the y-axis, and is mirror symmetric in both the xz- and yz-planes. The planar molecule BCl_3 (Fig. 48) does not change when rotated at 120° or 240° around the z-axis.

These symmetry properties are manifested in the normal oscillations of the molecules as well as in their spectra of radiation, absorption, and Raman scattering of light. To investigate these properties, we use group theory to systematically study the symmetry properties of various systems. Here, we provide proof of several useful properties of such systems with simple symmetry.

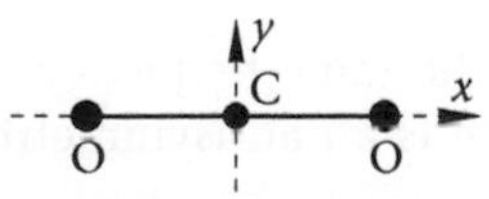

Figure 46 Molecule CO_2

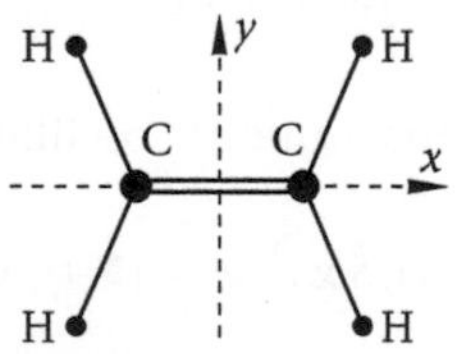

Figure 47 Molecule C_2H_4

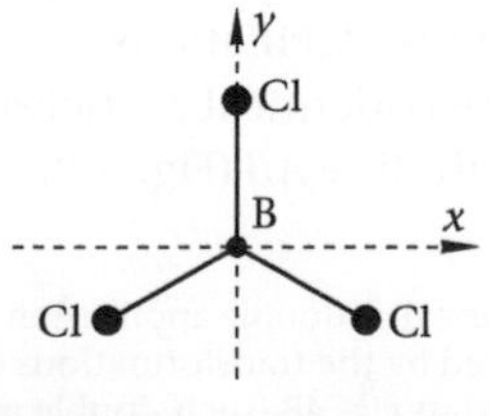

Figure 48 Molecule BCl_3

Let a system that performs linear oscillations (and, consequently, its Lagrangian $L(\mathbf{x}, \dot{\mathbf{x}}, t)$) not change its form under replacement

$$\mathbf{x} \to \hat{S}\mathbf{x}, \quad \dot{\mathbf{x}} \to \hat{S}\dot{\mathbf{x}}, \tag{28.1}$$

where the constant coefficients S_{ij} forming the matrix $\hat{S}$, satisfy the conditions[5]

$$\hat{S} = \hat{S}^T, \quad \hat{S}\hat{S} = \hat{E}. \tag{28.2}$$

We claim that the system has S symmetry and will show that the normal oscillations of such a system also exhibit certain symmetry properties.

Let $\mathbf{x} = \mathbf{A}\cos(\omega t + \varphi)$ be some normal oscillation. Since the replacement $\mathbf{x} \to \hat{S}\mathbf{x}$ does not change the Lagrangian or the equations of motion, $\hat{S}\mathbf{x}$ is also a normal oscillation with the same frequency. Thus, we can conclude the following:

a) If the frequency is not degenerate, then $\hat{S}\mathbf{x}$ should coincide with $\mathbf{x}$ except for a common factor. In other words, $\hat{S}\mathbf{x} = \lambda\mathbf{x}$. Since $\hat{S}\hat{S} = \hat{E}$, we obtain

$$\hat{S}\hat{S}\mathbf{x} = \lambda\hat{S}\mathbf{x} = \lambda^2\mathbf{x} = \mathbf{x}$$

or $\lambda = \pm 1$. Thus, the oscillation $\mathbf{x} = \mathbf{A}\cos(\omega t + \varphi)$ whose frequency is not degenerate is either symmetric, i.e. $\hat{S}\mathbf{A} = \mathbf{A}$, or antisymmetric, i.e. $\hat{S}\mathbf{A} = -\mathbf{A}$, with respect to the transformation S;

b) If the frequency ω is degenerate, then $\hat{S}\mathbf{x}$ can differ from and need not coincide with $\mathbf{x}$. However, the sum and difference of these two solutions

$$\mathbf{x} \pm \hat{S}\mathbf{x} = (\mathbf{A} \pm \hat{S}\mathbf{A})\cos(\omega t + \varphi)$$

are also normal oscillations with the same frequency. Moreover, the sum is a symmetric oscillation, while the difference is an antisymmetric oscillation with respect to the transformation S;

c) Suppose that an external force $\mathbf{F}(t)$ acts on the system such that

$$\hat{S}\mathbf{F}(t) = +\mathbf{F}(t).$$

If the normal oscillation $\mathbf{x}_a$ is an antisymmetric oscillation, i.e. $\hat{S}\mathbf{x}_a = -\mathbf{x}_a$, then

$$\left(\hat{S}\,\mathbf{F}(t), \hat{S}\,\mathbf{x}_a\right) = -\left(\mathbf{F}(t), \mathbf{x}_a\right)$$

due to properties (2). Hence, we obtain $(\mathbf{F}(t), \mathbf{x}_a) = 0$. In other words, the projection of the force onto a given oscillation is zero, and therefore the symmetric force does not affect antisymmetric oscillations (cf. § 25). Similarly, we can demonstrate that the antisymmetric force does not influence the symmetric oscillations.

A simple example. Consider two identical particles connected with identical springs that can only move along the straight line AB (Fig. 49). Let x_1 and x_2 be the displacements

[5] Condition $\hat{S}\hat{S} = \hat{E}$ or $\sum_j S_{ij}S_{jk} = \delta_{ik}$ means that double application of such a replacement returns the system to its original state. This property is possessed by the transformations considered above for the examples in Fig. 46 and 47. In contrast, in the last example, in Fig. 48, such double application of the rotation over 120° will not return the system to its original state.

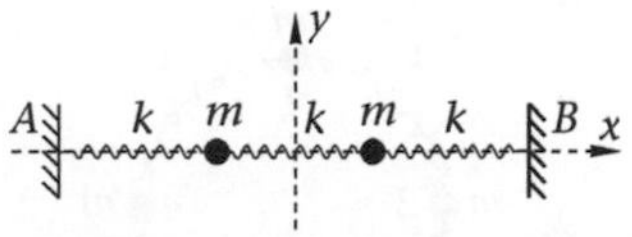

Figure 49 A simple symmetrical system

of the particles from their equilibrium positions. The system does not change when we turn it 180° around the y-axis, i.e. the Lagrangian of the system

$$L = \frac{1}{2}m(\dot{x}_1^2 + \dot{x}_2^2) - \frac{1}{2}k\left[x_1^2 + (x_1 - x_2)^2 + x_2^2\right] \tag{28.3}$$

does not change with respect to the replacement

$$\mathbf{x} = \begin{pmatrix} x_1 \\ x_2 \end{pmatrix} \to \hat{S}\mathbf{x} = \begin{pmatrix} -x_2 \\ -x_1 \end{pmatrix}, \quad \hat{S} = \begin{pmatrix} 0 & -1 \\ -1 & 0 \end{pmatrix}. \tag{28.4}$$

It is easy to verify that the matrix $\hat{S}$ satisfies conditions (2). Therefore, all oscillations of this system will be either symmetric, $\hat{S}\mathbf{x}_s = +\mathbf{x}_s$, or antisymmetric, $\hat{S}\mathbf{x}_a = -\mathbf{x}_a$. It is sufficient to adhere to these requirements to find the forms of normal oscillations (Fig. 50):

$$\mathbf{x}_s = \begin{pmatrix} 1 \\ -1 \end{pmatrix} a_s \cos(\omega_s t + \varphi_s),$$

$$\mathbf{x}_a = \begin{pmatrix} 1 \\ 1 \end{pmatrix} a_a \cos(\omega_a t + \varphi_a).$$

Suppose now that points A and B move simultaneously according to the law $x_A = x_B = b\cos(\gamma t + \varphi)$. This means that the external force acting upon this system is

$$\mathbf{F}(t) = \begin{pmatrix} 1 \\ 1 \end{pmatrix} kb\cos(\gamma t + \varphi).$$

This force is antisymmetric: $\hat{S}\mathbf{F}(t) = -\mathbf{F}(t)$; the vector of the force is directed along $\mathbf{x}_a$, and orthogonal to $\mathbf{x}_s$. Therefore, this force does not affect the symmetrical oscillation and causes the resonant amplification only in the antisymmetric one (at $\gamma \to \omega_a$).

More complex and meaningful examples regarding oscillations of molecules shown in Figs. 46–48 can be found in [3], problems 6.49, 6.51, and 6.52.

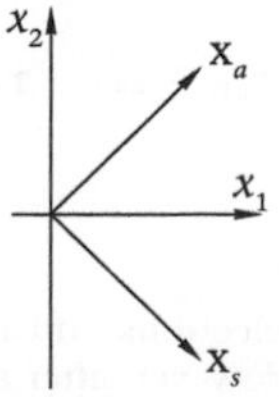

Figure 50 Vectors of normal oscillations of the system shown in Fig. 49

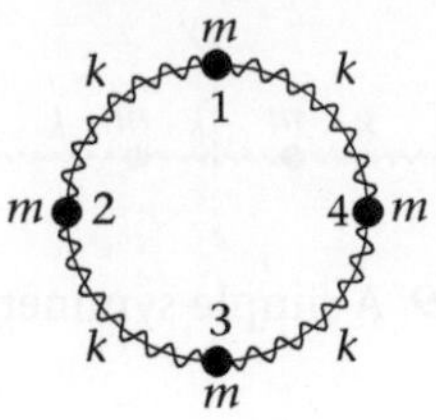

Figure 51 To Problem 3.1

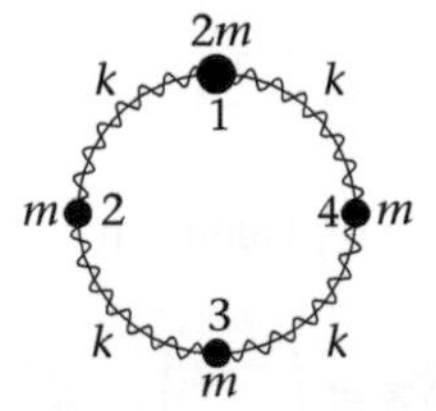

Figure 52 To problem 3.2

Problems

28.1. Find the normal coordinates of the system of four identical particles on a ring (Fig. 51).

Hint: it is convenient to use the properties of symmetry and mutual orthogonality of the normal oscillations.

28.2. Find the normal oscillations of a system of four particles on ring (Fig. 52).

§ 29 Oscillations of molecules

A molecule of N atoms possesses $3N$ degrees of freedom, corresponding to the $3N$ components of the radius vectors of the atoms: $\mathbf{r}_1, , \mathbf{r}_2, , \ldots, , \mathbf{r}_N$. When computing the normal oscillations of the molecule, both translational motion and rotation of the molecule as a whole must be excluded.[6] The translational motion of a molecule can be eliminated by switching to the system of the centre of mass, in which the atomic coordinates are bound by the condition

$$\sum_{a=1}^{N} m_a \mathbf{r}_a = 0\,, \tag{29.1}$$

where m_a is the mass of the a-th atom. Suppose that $\mathbf{u}_a$ represents the displacement of the a-th atom from its equilibrium position determined by the radius vector $\mathbf{r}_{a0}$, then

$$\mathbf{r}_a = \mathbf{r}_{a0} + \mathbf{u}_a\,, \quad \dot{\mathbf{r}}_a = \dot{\mathbf{u}}_a\,. \tag{29.2}$$

Since in a state of equilibrium,

[6] It should be noted that the motion of electrons and nuclei in a molecule is described not by classical mechanics but by quantum mechanics. However, after averaging over the fast motion of electrons, the oscillatory motion of nuclei can be considered classical in a certain approximation.

$$\sum_{a=1}^{N} m_a \mathbf{r}_{a0} = 0\,,$$

the above condition can be rewritten in the following form

$$\sum_{a=1}^{N} m_a \mathbf{u}_a = 0\,. \tag{29.3}$$

The same condition can be obtained by utilizing the known relations of orthogonality of normal oscillations (see § 24). Indeed, small oscillations of a molecule are described by the Lagrangian

$$L(\mathbf{u}, \dot{\mathbf{u}}) = \frac{1}{2}\sum_{a=1}^{N} m_a \dot{\mathbf{u}}_a^2 - \frac{1}{2}\sum_{a,b=1}^{N} \left(\mathbf{u}_a, \hat{k}_{ab}\mathbf{u}_b\right)\,, \tag{29.4}$$

where $\hat{k}_{ab}$ is the stiffness matrix. Among the solutions of the Lagrangian equations

$$m_a \ddot{\mathbf{u}}_a + \sum_{b=1}^{N} \hat{k}_{ab}\mathbf{u}_b = 0 \tag{29.5}$$

there are solutions

$$\mathbf{x}^{(\alpha)} = (\mathbf{u}_1, \dots, \mathbf{u}_N) \tag{29.6}$$

corresponding to normal oscillations with non-zero frequencies ω_α. Apart from those, there also exist solutions with frequency $\omega_0 = 0$ corresponding to the motion of the molecule as a whole with velocity $\mathbf{V}$:

$$\mathbf{x}^{(0)} = (\mathbf{V}t, \dots, \mathbf{V}t)\,. \tag{29.7}$$

The orthogonality relation for solutions (6) and (7) takes the form (in the mass metric (24.5*a*)):

$$\mathbf{V}t \cdot \sum_{a=1}^{N} m_a \mathbf{u}_a = 0\,.$$

From the arbitrariness of the vector $\mathbf{V}$, it follows that Eq. (3) is satisfied.

To eliminate the rotation of the molecule as a whole, we can ensure that its angular momentum is zero. Alternatively, we can seek a solution for Eq. (5) that corresponds to a small rotation of the molecule as a whole at an angle $\boldsymbol{\varepsilon} = \boldsymbol{\Omega}\,\delta t$. This is achieved by using the following expression:

$$\mathbf{x}^{(0)} = ([\boldsymbol{\varepsilon}, \mathbf{r}_{10}], \dots, [\boldsymbol{\varepsilon}, \mathbf{r}_{n0}])\,. \tag{29.8}$$

We can then use the orthogonality relation (24.5a) to obtain a solution for normal oscillation (6):

$$\sum_{a=1}^{N} [\boldsymbol{\varepsilon}, \mathbf{r}_{a0}] \cdot m_a \mathbf{u}_a = 0$$

which can be rewritten as

$$\boldsymbol{\varepsilon} \cdot \sum_{a=1}^{N} m_a \left[\mathbf{r}_{a0}, \mathbf{u}_a\right] = 0 \,.$$

Since the vector $\boldsymbol{\varepsilon}$ can be arbitrary, we find the condition

$$\sum_{a=1}^{N} m_a \left[\mathbf{r}_{a0}, \mathbf{u}_a\right] = 0 \tag{29.9}$$

for eliminating the rotation of the molecule as a whole.

For any molecule, except for linear ones, the two vector conditions (3) and (9) correspond to six ideal holonomic constraints. Therefore, the number of normal oscillations (with non-zero frequencies) of such molecules is equal to $3N - 6$. However, for a linear molecule with atoms arranged along the z-axis, the condition (9) leads to only two ideal holonomic constraints for the x and y components of the vectors $\mathbf{u}_a$. Therefore, the number of normal oscillations for a linear molecule is $3N - 5$.

Problems

29.1. Determine the normal oscillations of a linear symmetric CO_2 molecule, as shown in Fig. 46. Assume that the molecule's potential energy is solely dependent on the distances O—C and C—O, as well as the angle OCO.

29.2. Classify the eigen oscillations of an ethylene molecule (C_2H_4) based on their symmetry properties about the x- and y-axes, as shown in Fig. 47. In the equilibrium position, all atoms in the molecule lie in the same plane.

§ 30 Oscillations of linear chains

In the following sections, we will examine simple examples of chains of particles connected by springs, as they are the most elementary models utilized in solid state theory. The movement of atoms in a solid can be described using quantum mechanics. Nevertheless, concepts emerging from the resolution of problems related to classical chains are often quite useful in quantum theory as well. The electrical counterparts of such chains can be found in RF lines, employed in radio engineering.

30.1 *Equations of motion and boundary conditions*

In this study, we examine a discrete model of a one-dimensional stretched string that can undergo oscillations in the transverse plane. We consider N identical particles, each with a mass of m and connected by identical springs with an elastic constant of k. The particles are in equilibrium at a distance of l from each other, such that the coordinate of the n-th particle's equilibrium position is given by $X_n = n \cdot l$. The ends of the chain are fixed at points A and B (see Fig. 53). If the unstretched length of the spring is equal to l_0, then each spring's tension is equal to $f = k(l - l_0)$.

We consider small oscillations of the particles only in the y-axis direction, as any displacements in the x-axis direction are negligible. Letting y_n denote the n-th particle's

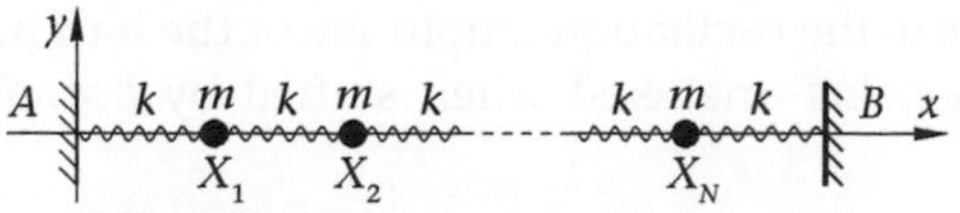

Figure 53 Chain with fixed ends

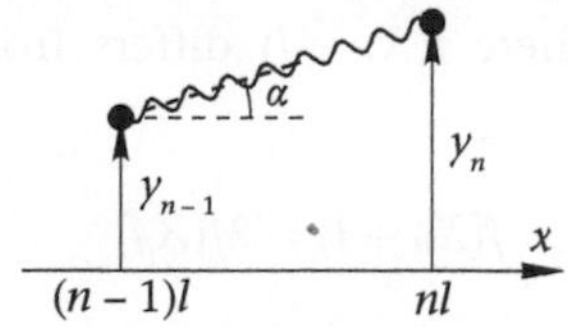

Figure 54 For the calculation of the force acting on the n-th particle from the side of the n-th spring

displacement from its equilibrium position along the y-axis, the restoring force on the n-th particle due to the n-th spring is given by

$$F_n = -f \sin\alpha = -f\frac{y_n - y_{n-1}}{l}$$

(see Fig. 54). Consequently, the Lagrangian governing the chain is expressed as

$$L = \frac{1}{2}m\sum_{n=1}^{N}\dot{y}_n^2 - \frac{f}{2l}\sum_{n=1}^{N+1}(y_n - y_{n-1})^2\,, \tag{30.1}$$

where y_0 and y_{N+1} serve as fictitious offsets for the ends of the chain and are set to

$$y_0 \equiv 0, \quad y_N \equiv 0\,. \tag{30.2}$$

The Lagrangian equations governing the system take the form

$$\ddot{y}_n + \omega_0^2\,(2y_n - y_{n-1} - y_{n+1}) = 0\,, \quad n = 1, 2, \ldots, N\,, \tag{30.3}$$

where we denote $\omega_0 = \sqrt{f/(ml)}$.

30.2 *Travelling waves*

The solution to the problem of oscillations in this system, according to the general rules outlined in § 23, is possible but quite cumbersome. Instead, it would be more convenient to use physical considerations to predict that the normal oscillations should be standing waves. Starting with the system of equations (3), the boundary conditions (2) can be ignored and the number of particles assumed to be unlimited. This system of equations (3) describes an infinite chain, making it easy to find solutions in the form of travelling waves.

The infinite chain has the following property: a shift in the numbering of particles does not change the system of equations. This symmetry allows for a solution in the form given by the equation

$$y_n(t) = \mathrm{Re}\left[e^{i\omega t} f(X_n)\right]\,, \quad X_n = n \cdot l\,, \tag{30.4a}$$

where $f(X_n)$ corresponds to the oscillation amplitude of the n-th particle. The Lagrangian of the infinite chain remains unaltered when shifted by l along the x-axis, thus the function

$$\operatorname{Re}\left[e^{i\omega t} f(X_n + l)\right] \tag{30.4b}$$

is also a harmonic solution, where $f(X_n + l)$ differs from $f(X_n)$ solely by a constant factor λ:

$$f(X_n + l) = \lambda f(X_n) \tag{30.5a}$$

and subsequently

$$f(X_n + l) = \lambda f(X_n) = \lambda^2 f(X_{n-1}) = \ldots = \lambda^n f(X_1)\,. \tag{30.5b}$$

Let us verify that such an assumption allows the derivation of a solution for an infinite chain. By substituting (4*a*) into Eq. (3) and taking into account (5), we can reduce the system of differential equations to one algebraic equation

$$\omega^2 = \omega_0^2 \left(2 - \lambda - \frac{1}{\lambda}\right), \tag{30.6}$$

which defines the relationship between ω and λ. Hence, we find

$$\lambda_{1,2} = d \pm \sqrt{d^2 - 1}\,, \quad d = 1 - \frac{\omega^2}{2\omega_0^2}.$$

Note that

$$\lambda_1 \lambda_2 = 1\,.$$

When $\omega < 2\omega_0$, $d < 1$ and the roots $\lambda_{1,2}$ are complex conjugates: $\lambda_1 = \lambda_2^*$ and $|\lambda_{1,2}| = 1$. This solution corresponds to travelling waves. For $\omega > 2\omega_0$, $d^2 > 1$ and the roots $\lambda_{1,2}$ are real and negative: $\lambda_{1,2} < 0$, with $|\lambda_1| > 1$ and $|\lambda_2| < 1$. This solution corresponds to oscillations, for which the oscillation amplitudes of particles increase (fall) along the chain.

Finally, for $|\lambda| = 1$, λ can be expressed as

$$\lambda = e^{\mp i\varphi}, \tag{30.7a}$$

which leads to

$$\omega^2 = 4\omega_0^2 \sin^2 \frac{\varphi}{2} \tag{30.8}$$

and

$$y_n = \operatorname{Re}\left[A e^{i(\omega t \mp n\varphi)}\right]. \tag{30.9a}$$

Introducing the notation $K = \varphi / l$, we obtain equation

$$\lambda = e^{\mp iKl} \tag{30.7b}$$

that describes waves that travel along or against the x-axis. The solutions are given by

$$y_n = \mathrm{Re}\left[Ae^{i(\omega t \mp KX_n)}\right]. \tag{30.9b}$$

where the frequency ω determines the period of oscillation in time as $T = 2\pi/\omega$, and the wave vector K defines the 'period' of oscillations in space – the wavelength being

$$\Lambda = \frac{2\pi}{K} = \frac{2\pi l}{\varphi}. \tag{30.10}$$

The quantity φ represents the difference in the phases of oscillations of neighbouring particles, as shown in Eq. (9). Eq. (8) describes the relationship between the frequency ω (or wave vector K) and this phase difference, known as the *law of dispersion*. From this equation, we can see that the frequencies of the travelling waves lie in the interval $0 < \omega < 2\omega_0$. The point of the constant phase $\omega t \mp KX_n = \mathrm{const}$ moves along the x-axis according to the law

$$X_n = \pm\frac{\omega}{K}t - \frac{\mathrm{const}}{K}, \tag{30.11}$$

where the positive sign corresponds to waves travelling to the right, while the negative sign corresponds to waves travelling to the left.

For the case of $\lambda < 0$, we can represent the value of λ as

$$\lambda = -e^{\pm\psi} = -e^{\mp\varkappa l}, \quad \psi = \varkappa l, \tag{30.12}$$

which yields

$$\omega^2 = 4\omega_0^2 \cosh^2(\psi/2), \tag{30.13}$$

where the oscillation amplitudes either decrease or increase (depending on the sign) with increasing X_n

$$y_n = \mathrm{Re}\left[(-1)^n A\, e^{i\omega t \mp \varkappa X_n}\right] = \mathrm{Re}\left[(-1)^n A\, e^{\mp n\psi} e^{i\omega t}\right]. \tag{30.14}$$

The frequencies of such solutions lie in the interval $\omega > 2\omega_0$. We can obtain solutions (13) and (14) from Eq. (8) and (9) with the formal replacement

$$\varphi \to \pi - i\psi. \tag{30.15}$$

Thus, we have found two types of oscillations in an infinite chain: Eqs. (7) and (8) describe travelling waves which we can use to obtain free oscillations of a chain with fixed ends, whereas solutions (13) and (14) will be needed in § 32 when studying forced oscillations.

There is one more characteristic of the system, namely, the flow of energy along the chain, which differs significantly for the two types of oscillations. The energy transferred from the $(n-1)$-th particle to the n-th particle in time dt is equal to the work done by the force $F_n = -m\omega_0^2(y_n - y_{n-1})$ acting from the side of the n-th spring to shift the n-th particle by the distance $dy_n = \dot{y}_n\, dt$:

$$dE = F_n\, dy_n = -m\omega_0^2 (y_n - y_{n-1})\, \dot{y}_n\, dt.$$

Therefore, the energy flux carried by the wave along the x-axis is equal to

$$\frac{dE}{dt} = -m\omega_0^2 (y_n - y_{n-1}) \dot{y}_n .$$

When averaged over the oscillation period, the energy flux for the solution in the form of the travelling wave (9) is

$$\left\langle \frac{dE}{dt} \right\rangle = \pm \frac{1}{2} m\omega_0^2 \omega |A|^2 \sin\varphi .$$

On the other hand, for solution (14), the average energy flux is equal to zero.

30.3 *Standing waves and spectrum*

Now, from the general consideration let us zoom onto the specific problem of a chain with fixed ends. The boundary conditions (2) can be satisfied by choosing the superposition of waves travelling in both directions:

$$y_n = A_{(+)} e^{i(\omega t + n\varphi)} + A_{(-)} e^{i(\omega t - n\varphi)} .$$

Condition $y_0 = 0$ gives $A_{(-)} = -A_{(+)}$ or

$$y_n = \mathrm{Re}\left[2iA_{(+)} \sin n\varphi \, e^{i\omega t}\right] = A \sin n\varphi \, \cos(\omega t + \chi) , \quad 2iA_{(+)} = A \, e^{i\chi} , \tag{30.16}$$

which represents standing waves whose amplitudes depend sinusoidally on the particle number n. Possible discrete values of frequencies are determined from the condition at the other end $y_{N+1} = 0$:

$$\sin(N+1)\varphi = 0.$$

The equation $\sin(N+1)\varphi = 0$ leads to N independent solutions

$$\varphi_\alpha = \frac{\pi\alpha}{N+1}, \quad \alpha = 1, 2, ..., N. \tag{30.17}$$

Indeed, the values $\alpha = 0$ and $\alpha = N+1$ give vanishing solutions $y_n = 0$. Moreover, for $\alpha = N+s$ at $s>1$, the phase $\varphi_{N+s} = 2\pi - \varphi_{N-s+2}$, that is, the solutions corresponding to $\alpha = N+s$ are expressed in terms of solutions corresponding to $\alpha = N-s+2$. We find N different frequencies from Eqs. (8) and (17) (see Fig. 55):

$$\omega_\alpha = 2\omega_0 \sin\frac{\pi\alpha}{2(N+1)} , \quad \alpha = 1, 2, \ldots, N. \tag{30.18}$$

As the number N increases, the number of eigenfrequencies increases as well, but they are all located in the interval $0 < \omega_\alpha < 2\omega_0$. This interval is called the *allowed band,* in contrast to the *forbidden band,* $\omega > 2\omega_0$.

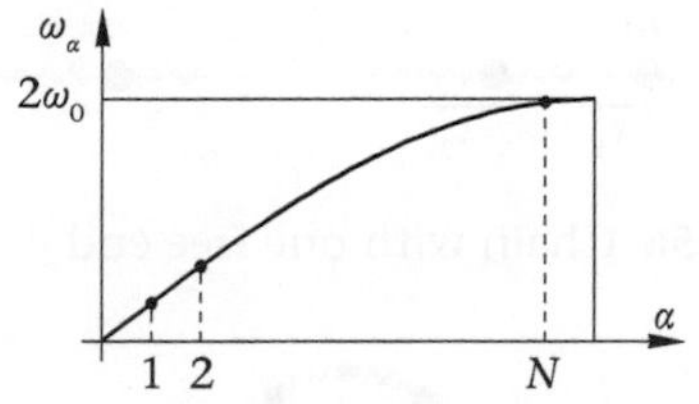

Figure 55 Frequency spectrum for the chain Fig. 42

The vector of the normal oscillation corresponding to the α-th frequency has the form

$$\mathbf{y}^{(\alpha)} = \text{const} \cdot \begin{pmatrix} \sin \varphi_\alpha \\ \sin 2\varphi_\alpha \\ \vdots \\ \sin N\varphi_\alpha \end{pmatrix} Q_\alpha(t); \quad Q_\alpha(t) = a_\alpha \cos(\omega_\alpha t + \chi_\alpha) . \tag{30.19}$$

The phase φ_α corresponds to the wavelength (10)

$$\Lambda_\alpha = \frac{2\pi}{K_\alpha} = \frac{2\pi l}{\varphi_\alpha} = \frac{2L}{\alpha}, \quad \alpha = 1, 2, ..., N, \tag{30.20}$$

where $L = l \cdot (N + 1)$ is the total length of the chain. Therefore, the α-th normal oscillation corresponds to such a standing wave in which the length of the chain fits exactly α half-waves.

If we choose in formula (19)

$$\text{const} = \frac{1}{\sqrt{\sum\limits_{n=1}^{N} \sin^2 n\varphi_\alpha}} = \sqrt{\frac{2}{N+1}},$$

then the various normal oscillations will be orthonormal:

$$(\mathbf{y}^{(\alpha)}, \mathbf{y}^{(\beta)}) = \delta_{\alpha\beta} Q_\alpha^2 .$$

The general solution is the superposition of all normal oscillations

$$y_n = \sum_{\alpha=1}^{N} U_{n\alpha} Q_\alpha(t) , \quad U_{n\alpha} = \sqrt{\frac{2}{N+1}} \sin \frac{\pi n \alpha}{N+1} . \tag{30.21}$$

The Lagrangian (1) in terms of the variables Q_α takes the form

$$L = \sum_{\alpha=1}^{N} L_\alpha; \quad L_\alpha = \frac{m}{2} \left(\dot{Q}_\alpha^2 - \omega_\alpha^2 Q_\alpha^2 \right) \tag{30.22}$$

which corresponds to a set of N different non-interacting oscillators.

Finally, we consider a chain of N identical particles of mass m connected by springs with elastic constant k unstretched in the equilibrium position. The particles are able to

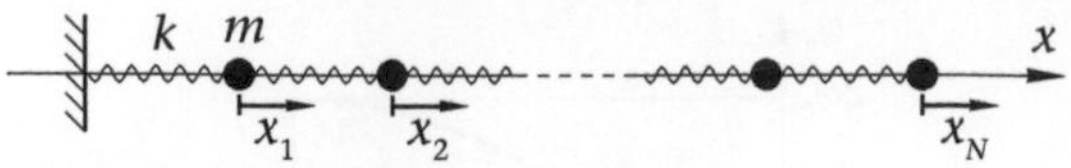

Figure 56 Chain with one free end

Figure 57 Closed chain on a ring

move only along the straight line AB as shown in Fig. 53. Let x_n denote the displacement of the n-th particle from the equilibrium position. It can be proven that the Lagrangian equations for this chain, subject to the additional conditions $x_0 = x_{N+1} \equiv 0$, take the form

$$\ddot{x}_n + \omega_0^2 \,(2x_n - x_{n-1} - x_{n+1}) = 0\,, \quad n = 1, 2, \ldots, N\,, \tag{30.23}$$

where $\omega_0 = \sqrt{k/m}$. These equations coincide with Eqs. (3) if one makes the replacements $y_n \to x_n$.

Problems

30.1. Determine the normal oscillations of a system of N identical particles with masses m connected by identical springs with elastic constant k and moving along a straight line with one free end as shown in Fig. 56.

30.2. Find the free oscillations of N identical particles with masses m which are connected by identical springs with elastic constant k and moving around the ring as shown in Fig. 57. Let the motion be that of a wave travelling around the ring. Check whether the energy flux equals the product of the linear energy density and the group velocity.

§ 31 The Born chain. Acoustic and optical oscillations of linear chains

The problem of oscillations in a chain with alternating particles of different masses, denoted by m and M respectively, is more complex. This chain consists of $2N$ particles, which are linked together by elastic springs of constant k. The particles move along the straight line AB, with chain ends A and B fixed (Fig. 58). Let x_n denote the displacement of the n-th particle from its equilibrium position. The Lagrangian equations governing this chain can be presented as follows:

$$m\ddot{x}_{2n-1} + k(2x_{2n-1} - x_{2n-2} - x_{2n}) = 0\,,$$
$$M\ddot{x}_{2n} + k(2x_{2n} - x_{2n-1} - x_{2n+1}) = 0,$$

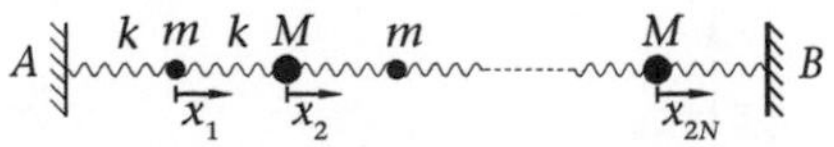

Figure 58 Chain of alternating particles with masses m and M

where $n = 1,, 2, \ldots N$. The boundary conditions are given by

$$x_0 = x_{2N+1} \equiv 0\,.$$

We are interested in finding a solution for standing waves with unique amplitudes for light and heavy particles, which can be represented by

$$\begin{aligned} x_{2n-1} =& A_\alpha \sin\left[(2n-1)\varphi_\alpha\right] \cos(\omega_\alpha t + \chi_\alpha)\,, \\ x_{2n} =& B_\alpha \sin\left[2n\varphi_\alpha\right] \cos(\omega_\alpha t + \chi_\alpha)\,. \end{aligned}$$

Substituting these expressions in the equations of motion, we obtain a system of two homogeneous linear algebraic equations for the amplitudes A_α and B_α. Solving these equations, we can establish a relationship between amplitudes A_α and B_α, as

$$B_\alpha = \frac{2k - m\omega_\alpha^2}{2k\cos\varphi_\alpha} A_\alpha\,, \quad \alpha = 1, 2, \ldots, N$$

and eigenfrequencies of oscillations, as

$$\omega^2_{(\mp)\alpha} = \frac{k}{\mu}\left(1 \mp \sqrt{1 - \frac{4\mu^2}{mM}\sin^2\varphi_\alpha}\right),$$

$$\mu = \frac{mM}{m+M}\,, \quad \varphi_\alpha = \frac{\pi\alpha}{2N+1}\,, \quad \alpha = 1, 2, \ldots N\,.$$

Two types of oscillations are distinguished in this case. The first type is known as *acoustic* oscillations, corresponding to low frequencies within the interval,

$$0 < \omega_{(-)\alpha} < \sqrt{\frac{2k}{M}}\,. \tag{31.1}$$

The second type, called *optical* oscillations, corresponds to higher frequencies within the interval

$$\sqrt{\frac{2k}{m}} < \omega_{(+)\alpha} < \sqrt{\frac{2k}{\mu}}\,, \tag{32.2}$$

(Fig. 59). These two intervals, denoted $(-)$ and $(+)$, respectively, constitute allowed bands that are separated by a forbidden band identified as

$$\sqrt{\frac{2k}{M}} < \omega < \sqrt{\frac{2k}{m}} \tag{33.3}$$

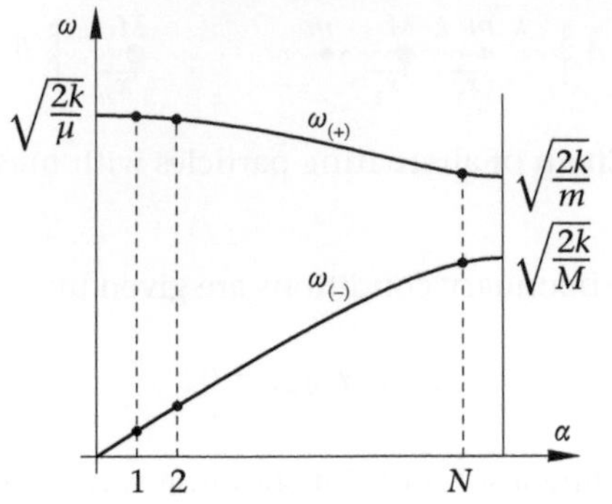

Figure 59 Frequency spectrum for the chain Fig. 58

which, for small values of $m \ll M$, features a substantial gap. Note that the high frequency region

$$\omega > \sqrt{\frac{2k}{\mu}}$$

is also a forbidden band.

It is also noteworthy that the amplitudes $A_{(-)\alpha}$ and $B_{(-)\alpha}$, which correspond to acoustic frequencies, exhibit the same signs. Therefore, neighbouring particles with masses m and M, i.e. $(2n-1)$-th and $(2n)$-th particles, oscillate in phase. In contrast, the amplitudes $A_{(+)\alpha}$ and $B_{(+)\alpha}$, corresponding to optical frequencies, exhibit opposite signs, indicating antiphase oscillations of neighbouring particles.

Figure 60 shows the distribution of oscillation amplitudes for $N=10$, $\alpha=2$, and $M=2m$, where the ordinate and the abscissa represent the numbers of particles and their corresponding amplitudes, respectively. Figure 60*a* depicts an acoustic mode, whereas Fig. 60b represents an optical mode. It is pertinent to note that the amplitudes $A_{(-)\alpha}$ and $B_{(-)\alpha}$ for acoustic oscillations exhibit very slight differences, nearly fitting on the same curve in Fig. 60*a*.

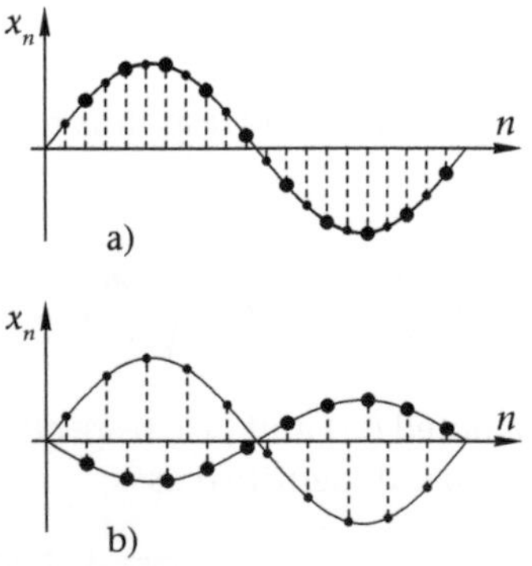

Figure 60 Distribution of amplitudes for oscillations of chain Fig. 58 at $N=10$, $\alpha=2$, $M=2m$: *a* – for acoustic; *b* – for optical modes

§ 32 Forced oscillations of linear chains under the action of a harmonic force

Let us now consider the forced oscillations of the chain illustrated in Fig. 53 that occur when the right end of the chain, i.e. point B, oscillates along the y-axis according to the law

$$y_B = b\cos(\gamma t + \chi).$$

The equations of motion in this scenario retain the same form as Eq. (30.3). However, the boundary conditions (30.2) are replaced with the conditions

$$y_0 \equiv 0\,, \quad y_{N+1} = b\cos(\gamma t + \chi)\,. \tag{32.1}$$

For $\gamma < 2\omega_0$ (within the allowed band), the solution can be represented in the form:

$$y_n = B\sin n\varphi\cos(\gamma t + \chi) \tag{32.2}$$

which is analogous to (30.9). This solution satisfies the boundary condition $y_0 = 0$ and the equations of motion (30.3) only if the phase φ satisfies the equation:

$$\gamma^2 = 4\omega_0^2\sin^2(\varphi/2) \tag{32.3}$$

which is analogous to (30.8). To satisfy the condition at the right end, it is necessary to fulfil the equation:

$$B\sin(N+1)\varphi = b\,,$$

which gives the final result for forced oscillations:

$$y_n = b\,\frac{\sin n\varphi}{\sin(N+1)\varphi}\cos(\gamma t + \chi)\,. \tag{32.4}$$

When $\gamma \ll \omega_0$, Eq. (3) provides $\varphi \approx \gamma/\omega_0 \ll 1$, and thus,

$$y_n \approx n\frac{b}{N+1}\cos(\gamma t + \chi)\,,$$

which means that all particles oscillate in phase while the oscillation amplitudes linearly increase with increasing particle number. When γ approaches ω_α, where ω_α is one of the eigen frequencies (30.18), φ approaches $\varphi_\alpha = \pi\alpha/(N+1)$, and the amplitudes of particle oscillations increase to infinity as $\sin(N+1)\varphi \to \sin(N+1)\varphi_\alpha = 0$. Therefore, resonances arise at each of the eigen frequencies in this chain.

For $\gamma > 2\omega_0$ (within the forbidden band), the solution can be represented as a superposition of solutions of the type (30.14):

$$y_n = (-1)^n (Ae^{n\psi} + Be^{-n\psi}) \cos(\gamma t + \chi) , \tag{32.5}$$

which satisfies the equations of motion (30.3) if

$$\gamma^2 = 4\omega_0^2 \cosh^2(\psi/2) . \tag{32.6}$$

The condition $y_0 = 0$ gives $A = -B$ or

$$y_n = (-1)^n 2A \sinh n\psi \cos(\gamma t + \chi) ,$$

while the condition at the other end

$$(-1)^{N+1} 2A \sinh(N+1)\psi = b$$

gives

$$y_n = (-1)^{N+1-n} b \frac{\sinh n\psi}{\sinh(N+1)\psi} \cos(\gamma t + \chi) . \tag{32.7}$$

In fact, this solution can be obtained directly from solution (4) with the substitution (30.15). The oscillation amplitudes decrease towards the left end of the chain, and each particle oscillates in antiphase with neighbouring particles. At $\gamma \gg 2\omega_0$, we have $\gamma \approx \omega_0 e^{-\psi}$ and

$$y_n = \frac{b}{(-\gamma^2/\omega_0^2)^{N+1-n}} \cos(\gamma t + \chi) ,$$

which means that the oscillation amplitudes exponentially decrease towards the left end of the chain. Taking into account (25.4a), this result is reasonable. For $\gamma \gg 2\omega_0$ (and hence $\gamma \gg \omega_\alpha$), the particle on the extreme right oscillates with a small amplitude in antiphase with the applied force of high frequency, while the $(N-1)$-th particle is essentially at rest. We can then consider the motion of the $(N-1)$-th particle as a forced oscillation caused by an applied force of high frequency arising from the N-th particle, and so on.

Note that long chains of capacitors and inductors are described by the same equations as the mechanical chains discussed above. From a radio engineering perspective, these chains can be considered as bandpass filters that transfer energy from one end of the chain to another quite efficiently within the area of allowed frequencies but fail to do so in the area of forbidden frequencies.

§ 33 Non-linear oscillations. Anharmonic corrections

Deviation from the quadratic approximation in potential energy becomes significant as the oscillation amplitudes increase, leading to non-linear oscillations. The field of non-linear oscillations is quite vast, but this course will only cover some undermentioned cases that typify the phenomena.

33.1 *One-dimensional non-linear oscillations*

Let us consider the simplest features of non-linear oscillations in the case of one-dimensional motion in a potential field. Taking into account the corrections to potential energy up to the third and fourth orders of smallness in deviation from the equilibrium position, we can write the Lagrangian function as follows:

$$L(x,\dot{x}) = \frac{1}{2}\left(m\dot{x}^2 - kx^2\right) - \frac{1}{3}m\alpha x^3 - \frac{1}{4}m\beta x^4,$$

where α and β are small constants. The coefficients are presented in this manner to simplify the equation of motion:

$$\ddot{x} + \omega_0^2 x = -\alpha x^2 - \beta x^3\,, \tag{33.1}$$

where $\omega_0^2 = k/m$. The solution to this equation is a periodic function of time with period $T = 2\pi/\omega$, dependent on energy. Although the exact solution (see (1.4)) can be expressed through elliptic functions, an easier, approximate solution may be useful for small oscillation amplitudes. To obtain it, we utilize the fact that the solution is a periodic function of time which can be written as a Fourier series:

$$x(t) = \sum_{n=0}^{\infty} a_n \cos(n\omega t)\,. \tag{33.2}$$

In this equation, the reference time is chosen such that the deviation of x is at its extremum for $t = 0$. The expansion coefficients a_n and frequency ω can be obtained from the equations of motion for a given energy. However, it is convenient to consider the amplitude of the fundamental harmonic as an independent parameter, denoted as

$$a_1 \equiv a,$$

while expressing the remaining amplitudes and frequency through it. The difference between ω and ω_0, denoted as

$$\delta\omega = \omega - \omega_0,$$

depends on a and is known as the *non-linear shift of frequency*.

We can substitute the Fourier series solution into the equation of motion and express all powers of the cosine on the right side in terms of higher harmonics. By equating the coefficients of each harmonic on both sides of the equation, we obtain an infinite system of non-linear equations. Each equation includes an infinite number of unknown coefficients a_n and an unknown frequency ω.

It is natural to expect that for a small non-linearity, the solution would not differ much from harmonic oscillations. In this study, the method of *successive approximations* is used to obtain an approximate solution for this system. This involves reducing the solution of a non-linear problem to a sequential calculation of corrections of ever higher order in a. We will limit our calculations for the variable x up to the third order in a, inclusive, and the frequency shift up to the second order in a. As the first approximation, we assume harmonic oscillations:

$$x^{(1)} = a\cos\omega t.$$

Here, ω is an unknown frequency, which is expected to differ only slightly from ω_0 and will be determined in subsequent approximations.

To obtain the second approximation, we substitute $x^{(1)}$, with an unknown frequency, into the small right side of Eq. (1) and express degrees of the cosine through the cosines of the harmonics:

$$\cos^2 \omega t = \frac{1}{2} + \frac{1}{2} \cos 2\omega t\,, \quad \cos^3 \omega t = \frac{3}{4} \cos \omega t + \frac{1}{4} \cos 3\omega t\,.$$

This yields an equation with a given driving force, containing all harmonics up to and including the third:

$$\ddot{x} + \omega_0^2 x = -\alpha a^2 \left(1 + \cos 2\omega t\right) - \frac{1}{4} \beta a^3 \left(3 \cos \omega t + \cos 3\omega t\right)\,.$$

Therefore, the particular solution will be sought in the form of a sum of harmonics up to the third inclusive:

$$x^{(2)}(t) = a_0^{(2)} + a \cos \omega t + a_2^{(2)} \cos 2\omega t + a_3^{(2)} \cos 3\omega t\,.$$

To obtain the coefficients of the harmonics, we write out the conditions for equality of the coefficients for the same harmonics on the right and left sides of the equation. This leads to the following four equations:

$$\omega_0^2 a_0^{(2)} = -\frac{1}{2} \alpha a^2\,, \tag{33.3}$$

$$(\omega^2 - \omega_0^2) a = \frac{3}{4} \beta a^3\,, \tag{33.4}$$

$$(4\omega^2 - \omega_0^2) a_2^{(2)} = \frac{1}{2} \alpha a^2\,, \tag{33.5}$$

$$(9\omega^2 - \omega_0^2) a_3^{(2)} = \frac{1}{4} \beta a^3\,. \tag{33.6}$$

Equation (4) depends only on the unknown frequency ω. Assuming that the non-linear frequency shift is small, we have:

$$\omega^2 - \omega_0^2 = 2\omega_0(\omega - \omega_0)$$

Using this, we obtain:

$$\delta\omega^{(2)} = \omega - \omega_0 = \frac{3\beta a^2}{8\omega_0}\,.$$

Next, Eqs. (3), (5), and (6) are utilized to obtain the amplitudes of harmonics, neglecting the difference between ω_0 and ω frequencies:

$$a_0^{(2)} = -\frac{\alpha a^2}{2\omega_0^2}, \quad a_2^{(2)} = \frac{\alpha a^2}{6\omega_0^2}, \quad a_3^{(2)} = \frac{\beta a^3}{32\omega_0^2}\,. \tag{33.7}$$

Thus, in the second approximation, we obtained the amplitudes of the zero and second harmonics, the frequency correction of the second order in a, and the amplitude of the third harmonic of the third order in a.

The second approximation can be substituted into the non-linear part of the equations to obtain the third approximation. Continuing like this, we can find the amplitudes of increasingly higher harmonics with higher orders of smallness in a. However, it is crucial to note that new corrections of the third order in a will appear in the third approximation and must be taken into account. When the second approximation is substituted into the quadratic term of the equation, terms proportional to $a_0^{(2)}\, a\cos\omega t$ and $a_2^{(2)}\, a\cos 2\omega t\cos\omega t$ will appear, which are of order a^3. Considering these terms will change the equalities (4) and (6) and introduce another correction in the frequency:

$$\delta\omega^{(3)} = -\frac{5\alpha^2 a^2}{12\omega_0^3}$$

and the addition to the amplitude of the third harmonic

$$a_3^{(3)} = \frac{\alpha^2 a^3}{48\omega_0^4}\,.$$

It can be verified that one further iteration will not change the corrections of the second and third orders in a. Finally, incorporating all corrections of the second and third orders into a, we have:

$$x(t) = a\cos\omega t - \frac{\alpha a^2}{6\omega_0^2}(3 - \cos 2\omega t) + \frac{a^3}{16\omega_0^2}\left(\frac{\alpha^2}{3\omega_0^2} + \frac{\beta}{2}\right)\cos 3\omega t\,,$$

$$\delta\omega = \left(\frac{3\beta}{8\omega_0} - \frac{5\alpha^2}{12\omega_0^3}\right)a^2\,. \tag{33.8}$$

The conditions for the applicability of the obtained formulae can now be specified. From the requirement of smallness of the amplitudes of higher harmonics as compared to a and the smallness of the correction to frequency as compared to ω_0, the following inequalities are obtained:

$$\frac{\alpha a}{\omega_0^2} \ll 1, \quad \frac{\beta a^2}{\omega_0^2} \ll 1\,. \tag{33.9}$$

33.2 *Multidimensional non-linear oscillations. Combination frequencies*

In the one-dimensional case, the first (linear) approximation yields harmonic oscillations at the fundamental frequency. However, by considering non-linearity in the second approximation, a shift occurs from the equilibrium position, resulting in oscillations at the double frequency. When extending this program to multidimensional oscillations, we also obtain oscillations corresponding to the first harmonics with frequencies $\omega_1, , \ldots, , \omega_\alpha, , \ldots, , \omega_\beta, , \ldots, , \omega_s$ in the first approximation. Nevertheless, in the second approximation, new phenomena arise that are manifested as oscillations at so-called *combination frequencies* $|\omega_\alpha \pm \omega_\beta|$.

To demonstrate how non-linear corrections arise in the multidimensional case and their outcomes, we consider a pendulum of mass m on a spring with elastic constant k in the gravity field g (Fig. 61). We assume vertical plane oscillations and that the unstretched length of the spring is l_0, with length $l = l_0 + (mg/k)$ in the equilibrium position. Cartesian coordinates x and y are used to denote the deviation of the pendulum from its equilibrium position. We then expand the Lagrangian of the pendulum

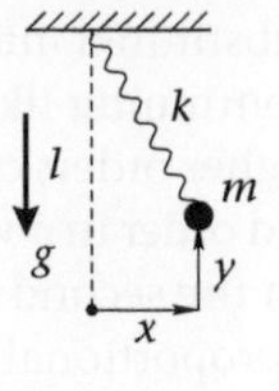

Figure 61 Spring pendulum

$$L = \frac{1}{2}m\left(\dot{x}^2 + \dot{y}^2\right) - \frac{1}{2}k\left[\sqrt{(l-y)^2 + x^2} - l_0\right]^2 - mgy$$

in a small deviation series up to the third order inclusive:

$$L = \frac{1}{2}m\left(\dot{x}^2 - \omega_x^2 x^2 + \dot{y}^2 - \omega_y^2 y^2 + 2\alpha x^2 y\right), \tag{33.10}$$

where the notations are used

$$\omega_x = \sqrt{\frac{g}{l}}, \quad \omega_y = \sqrt{\frac{k}{m}}, \quad \alpha = \frac{kl_0}{2ml^2}.$$

The equations of motion

$$\ddot{x} + \omega_x^2 x = 2\alpha xy,$$
$$\ddot{y} + \omega_y^2 y = \alpha x^2$$

are solved through the method of successive approximations:

$$x = x^{(1)} + x^{(2)} + \dots, \quad y = y^{(1)} + y^{(2)} + \dots.$$

As the first approximation, harmonic oscillations with frequencies ω_x and ω_y occur:

$$x^{(1)} = a\cos(\omega_x t + \varphi_x), \quad y^{(1)} = b\cos(\omega_y t + \varphi_y). \tag{33.11}$$

In the second order, the first equation yields the expression

$$\ddot{x}^{(2)} + \omega_x^2 x^{(2)} = 2\alpha\, x^{(1)} y^{(1)} = \alpha\, ab\left[\cos(\omega_+ t + \varphi_+) + \cos(\omega_- t + \varphi_-)\right],$$

where

$$\omega_\pm = \omega_y \pm \omega_x, \quad \varphi_\pm = \varphi_y \pm \varphi_x.$$

The solution to this equation is harmonic oscillations with combination frequencies $\omega_\pm$:

$$x^{(2)} = \frac{\alpha\, ab}{2\omega_y(2\omega_x + \omega_y)}\cos(\omega_+ t + \varphi_+) + \frac{\alpha\, ab}{2\omega_y(2\omega_x - \omega_y)}\cos(\omega_- t + \varphi_-). \tag{33.12}$$

Similarly, from the second equation we get

$$y^{(2)} = \frac{\alpha a^2}{2\omega_y^2} - \frac{\alpha a^2}{2(4\omega_x^2 - \omega_y^2)}\cos(2\omega_x t + 2\varphi_x), \tag{33.13}$$

Therefore, in the second order, the y coordinate undergoes a constant shift and oscillation with the doubled frequency $2\omega_x$.

These solutions are valid until the frequency ω_y is close to $2\omega_x$. When $\omega_y = 2\omega_x$, the anharmonic corrections are no longer small and can cause significant energy transfer from x- to y-oscillations and vice versa. This case is discussed in § 35.

§ 34 Non-linear resonances

The response of a harmonic oscillator to a periodic external force can be solved exactly and leads to the concept of resonance where the amplitude of forced oscillations sharply increases at a frequency close to the eigen oscillations of the oscillator. However, when accounting for the non-linear dependence of the restoring force on the deviation from the equilibrium position, two qualitatively new effects arise: the appearance of higher harmonics and a non-linear shift of frequencies. Accounting for the non-linearity of the equations of motion also changes the phenomenon of resonance. So, what exactly are these changes?

This problem does not have an exact solution, and therefore to obtain an approximate solution, we will employ the method of successive approximations outlined in § 33. To the equation for the non-linear oscillator (33.1), we add an external periodic force of the form $f(t) = f \cos \gamma t$ and weak friction $f_{\text{fr}} = -2m\lambda\dot{x}$, resulting in the equation:

$$\ddot{x} + \omega_0^2 x = \frac{f}{m} \cos \gamma t - \alpha x^2 - \beta x^3 - 2\lambda \dot{x} \,. \tag{34.1}$$

For the linear oscillator, the stable solution has the frequency of the external force and a shifted phase of $x = b\cos(\gamma t + \delta)$. For anharmonic forced oscillations, we will seek a solution in the form of a series:

$$x = b_0 + b\cos(\gamma t + \delta) + b_2 \cos 2(\gamma t + \delta) + b_3 \cos 3(\gamma t + \delta) + \dots \,. \tag{34.2}$$

As for the study of free oscillations of a non-linear oscillator without friction, we substitute the solution in the form of the series (2) into equation (1) and perform two iterations of successive approximations. For free oscillations of a non-linear oscillator, the amplitude a was given, and the amplitudes of the harmonics and the frequency shift of oscillations were calculated. In the case of forced oscillations, contrarily, what is given are the frequency γ and amplitude f of the external force, and it is the amplitude b of the fundamental harmonic that needs to be determined. The amplitudes of other harmonics can be found using similar formulae as before except that we replace ω_0 with γ.

The amplitude b of the fundamental harmonic and the phase shift δ are found from the equation obtained by equating contributions corresponding to oscillations on the fundamental frequency:

$$\left(-\gamma^2 + \omega^2\right) b\cos(\gamma t + \delta) - 2\lambda\gamma b \sin(\gamma t + \delta) = \frac{f}{m} \cos \gamma t \,, \tag{34.3}$$

where $\omega = \omega_0 + \delta\omega$ and $\delta\omega$ is the non-linear frequency shift of the eigen oscillations (33.8) proportional to the square of the amplitude. We then write:

$$\omega = \omega_0 + \varkappa b^2 \,, \quad \varkappa = \frac{3\beta}{8\omega_0} - \frac{5\alpha^2}{12\omega_0^3} \,. \tag{34.4}$$

Substituting

$$\cos\gamma t = \cos(\gamma t + \delta - \delta) = \cos(\gamma t + \delta)\cos\delta + \sin(\gamma t + \delta)\sin\delta$$

into the right side of Eq. (3) and equating the coefficients at $\cos(\gamma t + \delta)$ and $\sin(\gamma t + \delta)$ in the resulting equation, we get

$$(\omega^2 - \gamma^2)\, b = \frac{f}{m}\cos\delta, \quad -2\lambda\gamma b = \frac{f}{m}\sin\delta .$$

From here, one can find b and δ. The equation for b differs from the corresponding equation (26.4) in the linear case just by substituting

$$\omega_0^2 \to \omega^2 = (\omega_0 + \varkappa b^2)^2 \approx \omega_0^2 + 2\varkappa\omega_0 b^2 \tag{34.5}$$

and it is an equation of the third degree in b^2:

$$b^2\left[\left(\omega_0^2 + 2\varkappa\omega_0 b^2 - \gamma^2\right)^2 + 4\lambda^2\gamma^2\right] = \frac{f^2}{m^2}. \tag{34.6}$$

The position of the maximum of the curve $b(\gamma)$ and its value are defined by relations (26.5) and (26.6) with the change (5). Therefore, if $\varkappa$ is positive, we can expect that, compared to the linear case, the position of the maximum of the non-linear resonance will shift to the right, and its value will decrease. Conversely, if $\varkappa$ is negative, the position of the non-linear resonance maximum will shift to the left, and its value will increase. The dependence of $b(\gamma)$ for different values of λ and for a fixed positive value of $\varkappa$ is shown in Fig. 62 with bold lines. For comparison, the thin lines depict curves for the same values of λ and for $\varkappa = 0$. It can be seen that as λ decreases, not only does the maximum of the curves shift, but there are also kinks in the curves. In other words, a range of γ values arises where three different values of the amplitude $b(\gamma)$ correspond to a given frequency of the external force. Let us consider in more detail the resonance neighbourhood in the case of low friction $\lambda \ll \omega_0$. Denoting frequency detuning by

$$\epsilon = \gamma - \omega_0$$

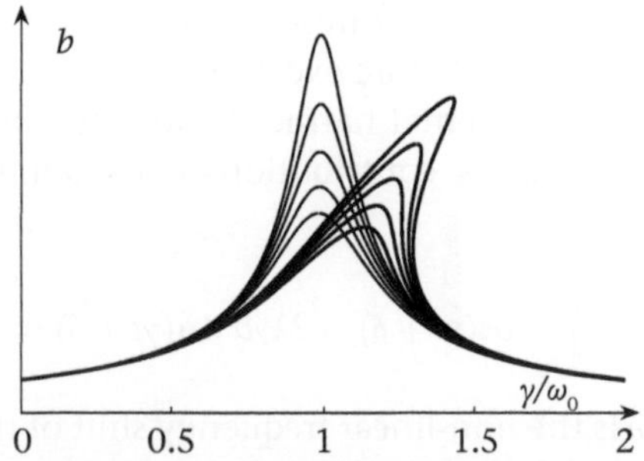

Figure 62 Dependence of the amplitudes of linear (thin curves) and non-linear (thick curves) oscillations b on the frequency of driving force γ for different values of the damping coefficient λ

and using approximate relations (26.7), we obtain

$$b^2[(\epsilon - \varkappa b^2)^2 + \lambda^2] = \frac{f^2}{4m^2\omega_0^2}. \tag{34.7}$$

This equation is cubic in b^2 and quadratic in ε. For each $b < b_{\max} \approx f/(2m\omega_0\lambda)$, there are two values of ε:

$$\epsilon = \varkappa b^2 \pm \sqrt{\left(\frac{f}{2m\omega_0 b}\right)^2 - \lambda^2}. \tag{34.8}$$

In the absence of non-linearity (at $\varkappa = 0$), the resonance curve coincides with that obtained in the linear case (Fig. 63*a*). It is easy to imagine how it deforms with the growth of $\varkappa$ at a constant force amplitude f. We assume $\varkappa > 0$. The maximum amplitude of oscillation remains approximately the same as in the linear case. However, in comparison to the linear case, for each pair of points corresponding to a given amplitude b, there is a shift to the right along the ε axis by $\varkappa b^2$, as depicted in Fig. 63*b*. The point corresponding to $b_{\max}$ experiences the largest offset. Thus, with the increase in $\varkappa$ on the right branch of the resonance curve $b(\epsilon)$, the tangent will become vertical at some point, which is also the inflection point. Qualitatively, the resonance becomes even stronger for a larger $\varkappa$. Now there exists an interval of values $\epsilon_1 < \epsilon < \epsilon_2$ in which a given ε corresponds to three values of oscillation amplitude, as shown in Fig. 63*c*.

We can determine the critical value $\varkappa_{\text{cr}}$ at which the transition occurs to this qualitatively new dependency. Equating the first and second derivative of the function $\epsilon(b^2)$ obtained from Eq. (8) with respect to b^2, we find that the tangent will be vertical at the

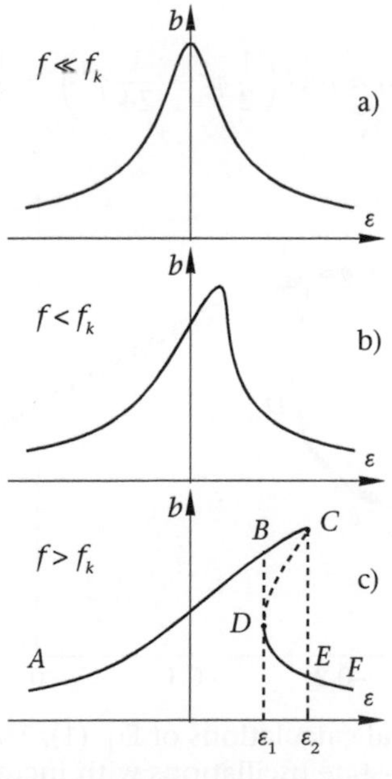

Figure 63 Dependence of the amplitude of the non-linear oscillations b on driving force frequency detuning $\epsilon = \gamma - \omega_0$ for different values of the non-linearity parameter $\varkappa$

point of inflection with coordinate $b^2 = 2\lambda/(\sqrt{3}\varkappa)$ under the condition:

$$\varkappa = \varkappa_{cr} \equiv \frac{32}{3\sqrt{3}} \frac{m^2\omega_0^2\lambda^3}{f^2}. \tag{34.9}$$

(Note that for negative $\varkappa$, the resonance maximum shifts to the left and the 'overturning' of the resonance curve occurs for $\varkappa < -\varkappa_{cr}$). If we consider the change in the resonance curve $b(\epsilon)$ with increasing amplitude of the external force f at a constant value of $\varkappa$, then the 'overturning' of the resonance curve will happen when:

$$f > f_{cr} = \sqrt{\frac{32}{3\sqrt{3}} \frac{m^2\omega_0^2\lambda^3}{|\varkappa|}}. \tag{34.10}$$

We will now follow the change in oscillation amplitude at slow adiabatic change in ε for the resonance curve *ABCDEF* shown in Fig. 63*c*. At increasing ε from the region of negative values, the amplitude of oscillations changes along the *ABC* branch, and at the point *C* it decreases in a jump to the value corresponding to the point *E* on the lower *DEF* branch (referred to as the 'breakdown' of oscillations). When ε is reduced from the range of positive values, the amplitude of oscillations varies along the *FED* line and increases abruptly to a value corresponding to the point *B* on the resonance curve (referred to as the 'hard excitation' of oscillations). Thus, the amplitude of oscillations in the interval $\epsilon_1 < \epsilon < \epsilon_2$ can have two different values depending on the history of the system. It can be shown that the third value of the amplitude corresponding to the *CD* branch is unstable, while the oscillations corresponding to the *ABC* and *FED* branches are stable.

In Fig. 64, the results of approximate analytical calculations are compared with numerical calculations of motion according to Eq. (1). The potential energy of a mathematical pendulum of length l at small angles $\varphi = x/l$ is used:

$$U = mgl(1 - \cos\varphi) \approx mgl\left(\frac{1}{2}\varphi^2 - \frac{1}{24}\varphi^4\right) = \frac{1}{2}m\omega_0^2x^2 + \frac{1}{4}m\beta x^4$$

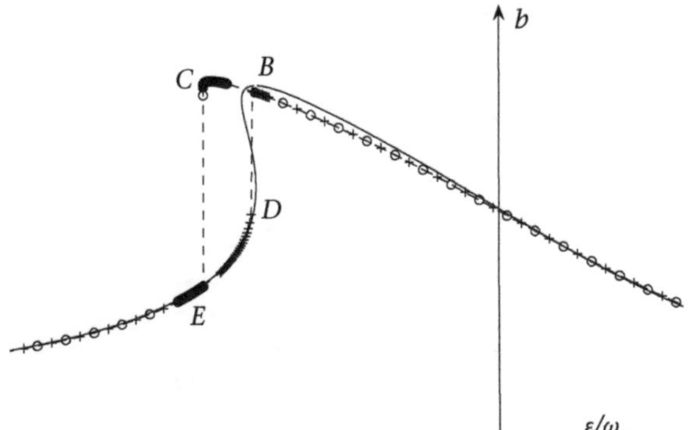

Figure 64 Results of numerical calculations of Eq. (1). The '+' symbol marks the values of the amplitude of the steady-state oscillations with increasing frequency of the external force γ, '–' symbol do the same with decreasing of γ. The step of the frequency is reduced around the frequency jump

with parameters

$$\alpha = 0, \ \beta = -\frac{\omega_0^2}{6l^3}, \ \lambda = 0.095\,\omega_0, \ f = 0.26\,m\omega_0^2 l.$$

A solid line shows the curve obtained by analytical calculations according to formula (7). The negative value of β corresponds to a decrease in the frequency of free oscillations of the pendulum with increasing amplitude. Therefore, the maximum of the curve lies at the negative value of ε. The crosses show the amplitudes of steady-state oscillations obtained as a result of the numerical solution of Eq. (1) with increasing frequency of the external force in very small steps. The zeros show the amplitudes obtained in the same way as the frequency decreases. When changing the frequency of the driving force, it was crucial that the phase of this force change without jumps. This ensured the passage along the resonance curve both in the direction of increasing frequency and in the direction of decreasing it, since the steady-state values of amplitude and phase of the oscillations $x(t)$ change little with frequency.

As a result, the jumps described in Fig. 64 were observed. It can also be seen that there is a difference between the results of the numerical calculation and the 'predictions' of the analytical calculations (7). The explanation for this discrepancy is that analytical calculations did not take into account the contributions of higher harmonics. It is not entirely correct to neglect them for $x \sim l$.

§ 35 Classical model of the Fermi resonance in a CO_2 molecule

The solutions found in § 33.2 hold true under the condition that the frequency ω_y remains sufficiently far from $2\omega_x$. However, when ω_y approaches $2\omega_x$, the anharmonic corrections become significant, potentially leading to a transfer of energy between x and y oscillations. In this section, we examine this scenario in greater detail. Note that this case is related to the coupling of longitudinal and flexural oscillations in a CO_2 molecule (known as *Fermi resonance*, as detailed in [13]) and the doubling and division of light frequency in non-linear optics (see [14]).

The Lagrangian function for this setup is given by

$$L = \frac{1}{2} m \left(\dot{x}^2 - \omega^2 x^2 + \dot{y}^2 - 4\omega^2 y^2 + 2\alpha x^2 y \right).$$

The equations of motion are

$$\ddot{x} + \omega^2 x = 2\alpha x y,$$
$$\ddot{y} + 4\omega^2 y = \alpha x^2.$$

We are looking for a solution in the form

$$x = Ae^{i\omega t} + A^* e^{-i\omega t} + \delta x,$$
$$y = Be^{2i\omega t} + B^* e^{-2i\omega t} + \delta y,$$

where $A = A(t)$ and $B = B(t)$ are slowly varying amplitudes of oscillations, while more rapidly oscillating terms δx and δy can be disregarded. Specifically,

$$|\ddot{A}| \ll \omega|\dot{A}| \ll \omega^2 |A|, \ |\ddot{B}| \ll \omega|\dot{B}| \ll \omega^2 |B|, \ \delta x \sim \delta y \ll |A|.$$

By only keeping the terms with $e^{i\omega t}$ (respectively $e^{2i\omega t}$) and disregarding $|\ddot{A}|$, $|\ddot{B}|$, we obtain

$$\dot{A} = -i\varepsilon BA^*\,, \tag{35.1}$$

$$4\dot{B} = -i\varepsilon A^2\,, \tag{35.2}$$

where the small parameter is introduced

$$\varepsilon = \frac{\alpha}{\omega}\,.$$

Two integrals of motion exist in this system. One corresponds to the conservation of energy:

$$|A|^2 + 4|B|^2 = C = \text{const}\,. \tag{35.3}$$

The constancy of C is easy to verify by direct differentiation in time using Eqs. (1)–(2). Indeed, on the right side of the equation

$$\frac{dC}{dt} = \dot{A}A^* + 4\dot{B}B^* + A\dot{A}^* + 4B\dot{B}^*$$

the first two terms can be found by multiplying the equation (1) on A^*, and equation (2) on B^* and adding up the results:

$$\dot{A}A^* + 4\dot{B}B^* = -i\varepsilon D\,,$$

where

$$D = A^{*2}B + A^2B^*\,. \tag{35.4}$$

The quantity D turns out to be real, and therefore

$$\frac{dC}{dt} = (-i\varepsilon D) + (-i\varepsilon D)^* = 0\,.$$

Proceeding in a similar manner, it is easy to prove that the second integral of motion is equivalent to the quantity D itself. By utilizing these two integrals of motion, one can determine the law of change with respect to time of the energy of the oscillation along the x-axis. This energy is proportional to the value

$$z(t) = |A(t)|^2\,.$$

Using Eq. (1), we obtain

$$\frac{dz}{dt} = i\varepsilon(A^{*2}B - A^2B^*)\,.$$

By squaring this equation and taking into account the integrals of motion (3)–(4), we arrive at

$$\dot{z}^2 = -\varepsilon^2\left[\left(A^{*2}B + A^2B^*\right)^2 - 4|A|^4|B|^2\right] = -\varepsilon^2\left[D^2 - |A|^4\left(C - |A|^2\right)\right]\,.$$

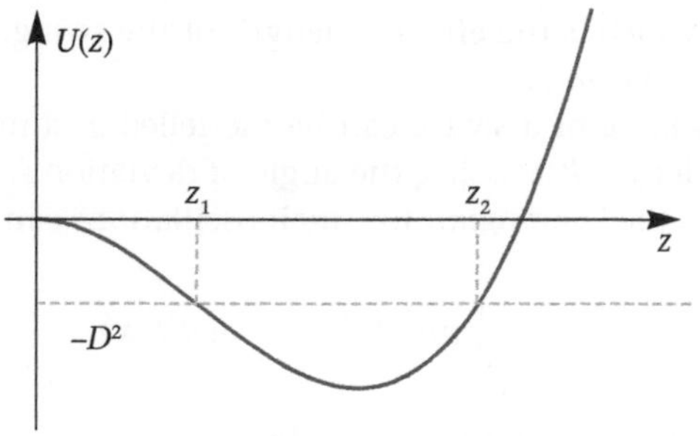

Figure 65 Determining the limits of change in the value $z = |A|^2$

This equation can be rewritten as

$$\dot{z}^2 + \varepsilon^2 U(z) = -\varepsilon^2 D^2\,, \quad U(z) = (z - C)\, z^2\,, \tag{35.5}$$

which is analogous to the law of conservation of energy for the problem of one-dimensional motion of a particle with coordinate $z = |A|^2$ and energy $-\varepsilon^2 D^2$ in the potential field $\varepsilon^2 U(z)$. This demonstrates that changes in the quantity $|A(t)|^2 = z(t)$ are best analysed by examining the graph $U(|A|^2) = (|A|^2 - C)|A|^4$ (Fig. 65). This graph shows that the amplitude $|A|$ undergoes oscillations within the bounds

$$z_1 \leq |A|^2 \leq z_2$$

which are determined by the equation

$$U(z_{1,2}) = -D^2\,.$$

In other words, in our system, beats occur with energy being pumped from oscillations along the x-axis into oscillations along the y-axis and vice versa. The period of these beats can be determined from the equation

$$T = \frac{2}{\varepsilon} \int_{z_1}^{z_2} \frac{dz}{\sqrt{-D^2 - U(z)}}\,.$$

The time dependence of the amplitudes $|A|$ and $|B|$ can be expressed using elliptic functions.

§ 36 Parametric resonance

Now, visualize yourself as a child on a swing. Do you recall how you used to pump a swing? This can be achieved by either pushing it in synchronization with its oscillations or by periodically squatting and straightening up. Obviously, in order to swing, a person needs to get up when the swing is in its lowest position, and squat when the swing is deflected as far as possible, so that the frequency of squats is doubled more than the frequency of the swing oscillations. In terms of mechanics, this is the use of the resonant change in the system parameter which is the effective length of the swing. The phenomenon itself is called the *parametric resonance*.

Mathematically, pumping of a swing can be modelled as a mass m pendulum with a periodically changing length $l(t)$. Using the angle of deviation from the vertical, denoted by x, as the coordinate, the Lagrangian for small oscillations can be written as (cf. (15.6))

$$L = \frac{1}{2}\, ml^2(t)\, \dot{x}^2 - \frac{1}{2}\, mgl(t)\, x^2$$

and the resulting equation of motion is given by

$$\frac{d}{dt}\left[l^2(t)\frac{dx}{dt}\right] + gl(t)x = 0.$$

By defining a new 'time' t' as $dt' = dt/l^2(t)$, using the notation $\omega^2(t') = gl^3(t)$, and removing the prime sign, the equation of motion can be written as

$$\frac{d^2x}{dt^2} + \omega^2(t)\, x = 0\,, \tag{36.1}$$

where the frequency $\omega(t)$ changes periodically in time.

The *Mathieu equation*, given by

$$\frac{d^2x}{dt^2} + \omega_0^2\,(1 + h\cos\gamma t)\, x = 0\,, \tag{36.2}$$

is used as a mathematical model for swinging a swing, with the parameter $\omega^2(t)$ depending on time in a harmonic law as given by

$$\omega^2(t) = \omega_0^2(1 + h\cos\gamma t)\,. \tag{36.3}$$

The most effective way to pump a swing is to stand up at the minimum and squat at the maximum deviation, which corresponds to approximately equal frequencies of $\gamma \approx 2\omega_0$.

The condition for the existence of an increasing solution of the Mathieu equation, assuming $h \ll 1$ and $\gamma = 2\omega_0 + \epsilon$, is considered in the presence of small detuning ϵ from the parametric resonance condition $\gamma = 2\omega_0$. The Mathieu equation can be rewritten as

$$\frac{d^2x}{dt^2} + \omega_0^2\, x = -h\omega_0^2 x\cos\gamma t\,. \tag{36.4}$$

Using the smallness of h, an approximate solution can be constructed by assuming a non-zero eigen oscillation at the initial moment (for example, $x = a\cos\omega_0 t$). The right side of Eq. (4) can then be written as in the form corresponding to the driving external force:

$$-h\omega_0^2 x\cos\gamma t = -\frac{1}{2}\, h\omega_0^2 a\,[\cos\,(\omega_0 + \epsilon)t + \cos\,(3\omega_0 + \epsilon)t]\,.$$

The first term on the right side of this equality can be understood as a resonant force (while the second one, containing the third harmonic, can be discarded). This results in a slow growth in the oscillation amplitude due to the smallness of h.

We will now seek a solution in the form

$$x(t) = a(t)\cos(\gamma t/2) + b(t)\sin(\gamma t/2)\,,$$

where $a(t)$ and $b(t)$ are slowly varying functions of time. Upon substitution of this expression into Eq. (4), we retain only the resonant terms on the right-hand side, while neglecting the small second derivatives of $\ddot{a}$ and $\ddot{b}$ relative to the former on the left-hand side. Ultimately, we are left with a system of equations

$$4\dot{a} + (2\epsilon + h\omega_0)\, b = 0\,,$$
$$4\dot{b} - (2\epsilon - h\omega_0)\, a = 0\,.$$

If $|\epsilon| < h\omega_0/2$, then the solution of this system is

$$a(t) = \alpha_1\left(C_1 e^{-st} + C_2 e^{st}\right),$$
$$b(t) = \alpha_2\left(C_1 e^{-st} - C_2 e^{st}\right),$$

where

$$s = \frac{1}{4}\sqrt{(h\omega_0)^2 - 4\epsilon^2}, \quad \alpha_{1,2} = \sqrt{h\omega_0 \pm 2\epsilon}\,, \qquad (36.5)$$

contains terms that grow exponentially with time. The solution corresponding to the parametric resonance has the form

$$x(t) = Ce^{st}\cos\left(\frac{1}{2}\gamma t - \varphi\right) + De^{-st}\cos\left(\frac{1}{2}\gamma t + \varphi\right), \qquad (36.6)$$

where $\tan\varphi = \alpha_1/\alpha_2$ (Fig. 66).

Thus, oscillations generally increase indefinitely, with the rate of growth characterized by the quantity s being small. If $|\epsilon| > h\omega_0/2$, then $s = \pm(i/4)\sqrt{4\epsilon^2 - (h\omega_0)^2}$ and the oscillation amplitude does not increase, but slowly oscillates over time (see Fig. 67).

So far, we have studied the main parametric resonance at the frequency $\gamma \approx 2\omega_0$. In the example of pumping a swing, this corresponds to the situation where the person squats in the high position of the swing and gets up in its lower position twice for one period of oscillation of the swing itself. It is with this method that energy is added to the system. However, it is also possible to pump the swing by squatting and standing up only once

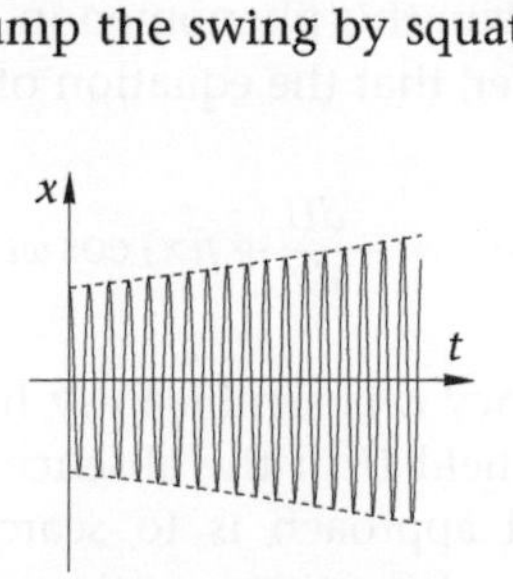

Figure 66 Parametric resonance

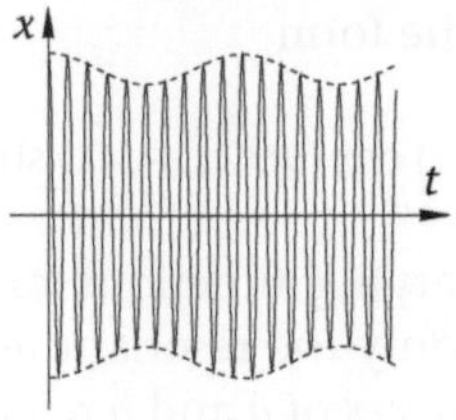

Figure 67 Beats

in the same period or in two swing periods, and so on. In other words, the parametric resonance can also occur at the frequency $\gamma \approx \omega_0/n$ for $n \geq 1$, although with less efficiency than at the main frequency $\gamma \approx 2\omega_0$ (i.e. in higher orders in h).

When friction is taken into account, the equation of motion becomes

$$\frac{d^2x}{dt^2} + \omega_0^2(1 + h\cos\gamma t)x + 2\lambda\frac{dx}{dt} = 0 \tag{36.7}$$

instead of Eq. (3). The solution of this equation, which increases with time, has the form (6) with

$$s = \frac{1}{4}\sqrt{(h\omega_0)^2 - 4\epsilon^2} - \lambda\,. \tag{36.8}$$

In contrast to the case without friction, the parametric buildup begins only for $h > 4\lambda/\omega_0$.

In real-world conditions, the increase in oscillation amplitude ceases when anharmonic corrections play a significant role, or when the oscillations themselves cause a reverse effect on the device which periodically changes the frequency (see [3], problem 8.7).

A more detailed and mathematically consistent analysis of this problem can be found in Supplement C.

§ 37 Motion in a rapidly oscillating field

As known, a flat pendulum has two equilibrium positions: the lower one, stable, and the upper unstable one. In the case of the *pendulum of P.L. Kapitsa*, the pendulum's suspension point oscillates vertically. Assuming the oscillation frequency is high enough, the pendulum's lower position's slow oscillation frequency changes, and the upper position may even become stable. To explain this phenomenon, the structure of the pendulum is not essential; it is crucial, however, that the equation of its motion is written as

$$m\ddot{x} = -\frac{dU}{dx} + f(x)\cos\omega t\,, \tag{37.1}$$

where the external force frequency ω is significantly higher than the characteristic frequencies of motion within the field U in the absence of an external force. With such frequency separation, a natural approach is to search for solutions in the form of $x(t) = X(t) + \xi(t)$, with $X(t)$ slow and averaging motion and $\xi(t) = \xi_0\cos\omega t$ high-frequency jitter with a small amplitude ξ_0. Small ξ_0 enables the use of a Taylor series expansion of the force and the transformation of the equation into the form

$$m\ddot{X} + m\ddot{\xi} = -\frac{dU(X)}{dX} + f(X)\cos\omega t - \frac{d^2U(X)}{dX^2}\xi + \frac{df(X)}{dX}\xi\cos\omega t\,. \tag{37.2}$$

Now, let us split this equation into two: high-frequency and low-frequency oscillations.

First, we obtain the high-frequency equation. On the right side of Eq. (2), the second and fourth terms oscillate with high frequency. Since it is expected – and will be demonstrated in a moment – that the amplitude ξ_0 is small, we can neglect the last term as it has the second-order smallness in amplitude and then obtain the equation

$$m\ddot{\xi} = f(X)\cos\omega t\,. \tag{37.3}$$

Its general solution contains a slowly varying solution of the homogeneous equation, which we neglect, and focus instead on the rapidly oscillating particular solution of an inhomogeneous equation, which we find by assuming slow coordinate $X(t)$ to be constant

$$\xi_0 = -\frac{f(X)}{m\omega^2}\,. \tag{37.4}$$

The equation for the slow variable is obtained by averaging Eq. (2) over the high-frequency oscillation's period. The average values of the linear terms on the right and left sides with respect to ξ and f are zero. However, the last term is not zero and adds an effective force, so that

$$m\ddot{X} = -\frac{dU}{dX} - \frac{1}{2m\omega^2}\frac{df(X)}{dX}f(X)\,. \tag{37.5}$$

The right side of the equation is written as $-dU_{\text{eff}}/dX$, where

$$U_{\text{eff}} = U + \frac{f^2}{4m\omega^2}\,, \tag{37.6}$$

which means that the average motion can be viewed as a field that mimics an additional field proportional to the square of the amplitude of the high-frequency force. This additional field pushes the particle to the region where the amplitude of the force is smaller.

A similar result holds for the case of three-dimensional motion. The slow and fast motion separation method is generally used in various mechanics-related problems.

Problem

37.1. Calculate the effective field for a pendulum, in which the suspension point undergoes vertical oscillations with an amplitude of a. Determine the condition under which the upper position of the pendulum becomes stable.

CHAPTER IV

Hamiltonian mechanics

§ 38 Hamiltonian equations

For a system with s degrees of freedom, there exist s Lagrangian equations given by

$$\frac{d}{dt}\frac{\partial L}{\partial \dot{q}_i} = \frac{\partial L}{\partial q_i}, \quad i = 1, 2, \dots, s, \tag{38.1}$$

which are second-order differential equations involving $\ddot{q}_i$, $\dot{q}_i$ and q_i. The generalized momenta are defined as

$$p_i = \frac{\partial L}{\partial \dot{q}_i} \tag{38.2}$$

and by virtue of (1), they satisfy the equation

$$\dot{p}_i = \frac{\partial L}{\partial q_i}. \tag{38.3}$$

The relations (2) and (3) demonstrate the possibility of transforming the second-order equations (1) to first-order equations for new variables q_i and p_i. Since the Lagrangian is a function of generalized coordinates, velocities and time, i.e. $L = L(q, \dot{q}, t)$, we can resolve relations (2) and (3) with respect to $\dot{q}_i$ and $\dot{p}_i$ as follows:

$$\dot{q}_i = f_i(q, p, t), \quad \dot{p}_i = g_i(q, p, t), \quad i = 1, 2, \dots, s, \tag{38.4}$$

where they can be represented as a system of $2s$ first-order equations with respect to the variables q_i and p_i.

38.1 *Hamiltonian function. Hamiltonian equations*

The explicit form of the functions f_i and g_i are obtained through the following process. The total differential of the Lagrangian function is given by

$$dL(q, \dot{q}, t) = \sum_{i=1}^{s}\left(\frac{\partial L}{\partial q_i}dq_i + \frac{\partial L}{\partial \dot{q}_i}d\dot{q}_i\right) + \frac{\partial L}{\partial t}dt.$$

Using (2) and (3), we can rewrite the above expression in terms of the differential of the generalized coordinates and momenta as

$$dL(q, \dot{q}, t) = \sum_{i=1}^{s}(\dot{p}_i dq_i + p_i d\dot{q}_i) + \frac{\partial L}{\partial t}dt. \tag{38.5}$$

Lectures on Analytical Mechanics. G. L. Kotkin, V. G. Serbo, A. I. Chernykh, Oxford University Press. © G. L. Kotkin, V. G. Serbo, A. I. Chernykh (2024). DOI: 10.1093/oso/9780198894674.003.0004

From here it is easy to see that for the function

$$H(q,p,t) = \sum_{i=1}^{s} \dot{q}_i p_i - L \tag{38.6}$$

the total differential reads

$$dH(q,p,t) = \sum_{i=1}^{s}(-\dot{p}_i dq_i + \dot{q}_i dp_i) - \frac{\partial L}{\partial t}dt. \tag{38.7}$$

Expression (6), considered as a function of generalized coordinates, momenta and time, is called the *Hamilton function* or *Hamiltonian*. Note that this quantity, taken as a function of time during the true motion of the system, is also known as the energy $E(t)$ (see § 16). We can express the differential of $H(q,p,t)$ in terms of partial derivatives, which allows us to obtain the desired *Hamiltonian* or *canonical* equations:

$$\dot{q}_i = \frac{\partial H(q,p,t)}{\partial p_i}, \quad \dot{p}_i = -\frac{\partial H(q,p,t)}{\partial q_i}, \quad i = 1, 2, \ldots, s \tag{38.8}$$

and equality

$$\frac{\partial H(q,p,t)}{\partial t} = -\frac{\partial L(q,\dot{q},t)}{\partial t}.$$

Here, the partial time derivative of the Lagrangian function is taken at constant coordinates and velocities, while for the Hamiltonian function it is taken at constant coordinates and momenta.

Unlike the Lagrangian equations (1), the Hamiltonian equations (8) are first-order equations, which also contain $2s$ unknown variables $q_1, \ldots, q_s, p_1, \ldots, p_s$ and their first time derivatives. For systems that are simple, the Lagrangian approach is often more convenient for analytical solutions. However, in many cases, when a general approach is needed (for example, in statistical physics), the Hamiltonian formalism is preferred. The special significance of Hamiltonian mechanics is due to the presence of a wider class of transformations in comparison to Lagrangian mechanics, with respect to which the Hamiltonian equations are covariant (see § 43). Finally, it is worth mentioning that the transition from the classical mechanics to quantum mechanics most naturally occurs within the framework of the Hamiltonian formalism.

The Hamiltonian equations, together with the Hamiltonian function, are results of the Lagrangian equations in mechanics. We want to emphasize the fact that the specific structure of the equations (8) provides several properties that hold true for any function $H(q,p,t)$ (see [2] and [4]).

Below there are some examples of Hamiltonians for different systems.

Example 1

Consider the Lagrangian of the form

$$L(q,\dot{q}) = \frac{1}{2}a(q)\dot{q}^2 + b(q)\dot{q} + c(q)$$

(cf. (16.6)). The quantity (6) is equal to

$$\dot{q}p - L = \frac{1}{2}a(q)\dot{q}^2 - c(q).$$

However, this quantity becomes a Hamiltonian function only after expressing the generalized velocity $\dot{q}$ in terms of the generalized momentum

$$p = \frac{\partial L}{\partial \dot{q}} = a(q)\dot{q} + b(q)$$

and the generalized coordinate:

$$\dot{q} = \frac{p - b(q)}{a(q)}.$$

As a result, the Hamiltonian is equal to

$$H(q, p, t) = \frac{[p - b(q)]^2}{2a(q)} - c(q). \tag{38.9}$$

For the case of a harmonic oscillator, with Lagrangian

$$L(x, \dot{x}) = \frac{1}{2}m\dot{x}^2 - \frac{1}{2}m\omega^2 x^2$$

the Hamiltonian is

$$H(x, p) = \frac{p^2}{2m} + \frac{1}{2}m\omega^2 x^2.$$

Example 2

For a particle under a central field in spherical coordinates where the Lagrangian is given by formula (10.8), the Hamiltonian is

$$H(r, \theta, \varphi, p_r, p_\theta, p_\varphi) = \frac{p_r^2}{2m} + \frac{p_\theta^2}{2mr^2} + \frac{p_\varphi^2}{2mr^2\sin^2\theta} + U(r). \tag{38.10}$$

Example 3

For a non-relativistic particle under an electromagnetic field which has the Lagrangian and energy provided in (12.4) and (16.8), respectively, and with the additional account of (12.5), we obtain

$$H(\mathbf{r}, \mathbf{p}, t) = \frac{1}{2m}\left(\mathbf{p} - \frac{e}{c}\mathbf{A}(\mathbf{r}, t)\right)^2 + e\phi(\mathbf{r}, t). \tag{38.11}$$

Example 4

The relativistic case is given by (14.1) and (14.3), where the Hamiltonian is

$$H(\mathbf{r}, \mathbf{p}, t) = \sqrt{\left(\mathbf{p} - \frac{e}{c}\mathbf{A}(\mathbf{r}, t)\right)^2 c^2 + m^2c^4} + e\phi(\mathbf{r}, t). \tag{38.12}$$

Example 5

A photon, i.e. a quantum of light, is a relativistic particle with zero mass, $m=0$, and charge, $e=0$. Its Hamiltonian for motion in vacuum is given by

$$H(\mathbf{r},\mathbf{p}) = c|\mathbf{p}|.$$

The propagation of light in a transparent isotropic medium with a refraction index of $n(\mathbf{r})$ is described, in the approximation of geometrical optics, by the Hamiltonian (cf. [9], § 85):

$$H(\mathbf{r},\mathbf{p}) = \frac{c}{n(\mathbf{r})}|\mathbf{p}|. \tag{38.12a}$$

The Hamiltonian equations take the form

$$\dot{\mathbf{r}} = \frac{c}{n}\frac{\mathbf{p}}{p}, \quad \dot{\mathbf{p}} = -\frac{cp}{n^2}\frac{\partial n}{\partial \mathbf{r}}, \quad p = |\mathbf{p}|.$$

In the context of geometric optics, the word 'particle' refers to a wave packet, and $\mathbf{r}(t)$ is the law of its motion, while $\dot{\mathbf{r}}$ is the group velocity. The vector $\mathbf{p}$ perpendicular to the wave front determines the wave vector of the electromagnetic wave.

38.2 *Integrals of motion in the Hamiltonian approach*

1. The easiest method to determine integrals of motion in the Hamiltonian approach is similar to using cyclic coordinates in the Lagrangian approach (see § 16). If the Hamiltonian does not depend on some generalized coordinate,

$$\frac{\partial H(q,p,t)}{\partial q_k} = 0,$$

then its corresponding generalized momentum is conserved:

$$p_k = \text{const.}$$

If the Hamilton function does not depend on some generalized momentum, its corresponding generalized coordinate is conserved.

2. Here is how the Hamiltonian changes as a system moves. Taking into account the equations of motion (8), we can obtain from (7):

$$\frac{dH(q,p,t)}{dt} = \sum_{i=1}^{s}\left(\dot{q}_i\frac{dp_i}{dt} - \dot{p}_i\frac{dq_i}{dt}\right) + \frac{\partial H}{\partial t} = \frac{\partial H}{\partial t}. \tag{38.13}$$

Therefore, if the Hamilton function does not explicitly depend on time,

$$\frac{\partial H(q,p,t)}{\partial t} = 0,$$

then the energy is conserved

$$E(t) = H(q(t),p(t)) = \text{const.}$$

Example 1

Let us consider in detail the solution of the Hamiltonian equations for a particle in a constant homogeneous magnetic field.[1] We direct the z-axis along the magnetic field $\mathbf{B} = (0, 0, B)$ and choose the vector potential in the form of

$$\mathbf{A} = (0, xB, 0). \tag{38.14}$$

The Hamiltonian function

$$H(x, y, z, p_x, p_y, p_z, t) = \frac{1}{2m}\left[p_x^2 + \left(p_y - \frac{e}{c}Bx\right)^2 + p_z^2\right] \tag{38.15}$$

does not depend on t, y, z, so the integrals of motion are energy and momentum components p_y and p_z:

$$p_y = m\dot{y} + \frac{eB}{c}x = \text{const}, \quad p_z = m\dot{z} = \text{const}.$$

It can be seen that the motion along the z-axis is uniform:

$$z = \frac{p_z}{m}t + z_0. \tag{38.16}$$

If we denote

$$\omega = \frac{eB}{mc}, \quad x_0 = \frac{cp_y}{eB} = \frac{p_y}{m\omega}, \tag{38.17}$$

then the Hamiltonian

$$H = \frac{p_x^2}{2m} + \frac{1}{2}m\omega^2(x - x_0)^2 + \frac{p_z^2}{2m} \tag{38.18}$$

coincides with the harmonic oscillator Hamiltonian (see Example 1 in the previous section) moving along the x-axis with the centre of oscillations located at the point x_0. It should be noted that x_0 and p_z are constant. Therefore, we can immediately write without using the equations of motion that:

$$x = x_0 + R\cos\omega(t - t_0), \quad p_x = m\dot{x} = -m\omega R\sin\omega(t - t_0), \tag{38.19}$$

where R and t_0 are constants of integration. From the equation

$$\dot{y} = \frac{\partial H}{\partial p_y} = m\omega^2(x_0 - x)\frac{\partial x_0}{\partial p_y} = -\omega R\cos\omega(t - t_0),$$

we find the dependence of $y(t)$ on t:

$$y = y_0 - R\sin\omega(t - t_0). \tag{38.20}$$

Thus, the particle moves along the z-axis at a constant velocity of p_z/m and rotates with an angular velocity of ω in the xy-plane. For $\omega > 0$, the rotation is clockwise around the circumference of radius R centred at the point with coordinates (x_0, y_0).

[1] The law of motion in this case is well-known. We will consider the solution of this problem in the Hamiltonian approach, which will be useful in quantum mechanics.

Note that the generalized momentum p_y has, up to a factor of $m\omega$, the meaning of the x-coordinate of the centre of the particle's orbit in the plane perpendicular to the direction of the magnetic field. For any other choice of the vector potential, the meaning of the generalized momenta will be different.

3. In case the Hamiltonian depends on the variables q_1 and p_1 only through the function $f(q_1, p_1)$, i.e.

$$H = H\left(f(q_1, p_1), q_2, p_2, \dots, q_s, p_s\right), \tag{38.21}$$

then $f(q_1, p_1)$ is an integral of motion.

Indeed,

$$\frac{df}{dt} = \frac{\partial f}{\partial q_1}\dot{q}_1 + \frac{\partial f}{\partial p_1}\dot{p}_1$$

and, using the Hamiltonian equations in the following form

$$\dot{q}_1 = \frac{\partial H}{\partial p_1} = \frac{\partial H}{\partial f}\frac{\partial f}{\partial p_1}, \quad \dot{p}_1 = -\frac{\partial H}{\partial q_1} = -\frac{\partial H}{\partial f}\frac{\partial f}{\partial q_1},$$

we get $df/dt = 0$. This implies that

$$f(q_1, p_1) = \text{const.} \tag{38.22}$$

The generalization of this result is obvious: if the Hamiltonian function has the form

$$H = H\left(f(q_1, p_1, \dots, q_k, p_k), q_{k+1}, p_{k+1}, \dots, q_s, p_s\right), \tag{38.23}$$

then $f(q_1, p_1, \dots, q_k, p_k)$ is the integral of motion.

Example 2

A particle moves in a dipole field described by (17.11). In the spherical coordinates with the z-axis directed along the dipole axis, the Hamiltonian takes the form (cf. (38.10)),

$$H = \frac{p_r^2}{2m} + \frac{1}{2mr^2}f(\theta, p_\theta, \varphi, p_\varphi), \quad f = p_\theta^2 + \frac{p_\varphi^2}{\sin^2\theta} + 2ma\cos\theta.$$

According to Example 1, the function f is the integral of motion. Taking into account Eq. (10.9), it can be rewritten as

$$f = \mathbf{M}^2 + 2ma\cos\theta. \tag{38.24}$$

In addition to this result, the integrals of motion include the energy E, the generalized momentum p_φ, and quantity (17.13).

Problems

38.1. Find the Hamiltonian of the anharmonic oscillator if the Lagrangian reads

$$L(x, \dot{x}) = \frac{1}{2}\left(m\dot{x}^2 - m\omega^2 x^2\right) - \alpha x^3 + \beta x \dot{x}^2.$$

38.2. Find the Lagrangian of a system if the Hamiltonian is

$$H(\mathbf{p}, \mathbf{r}) = \frac{\mathbf{p}^2}{2m} - \mathbf{pa},$$

where **a** is a constant vector.

38.3. Find the law of the particle motion with the Hamiltonian

$$H(p, x) = \frac{p^2}{2m} + \frac{1}{2} m\omega_0^2 x^2 + \lambda \left(\frac{p^2}{2m} + \frac{1}{2} m\omega_0^2 x^2\right)^2.$$

38.4. Find the equations of particle motion, if the Hamiltonian is

$$H(\mathbf{r}, \mathbf{p}) = \frac{c|\mathbf{p}|}{n(\mathbf{r})}$$

(that is a ray of light in a transparent medium with refractive index $n(\mathbf{r})$).
Find the trajectory of the particle if $n(\mathbf{r}) = az$.

38.5. Find the cross section for particles falling into the centre of the field

$$U(\mathbf{r}) = \frac{\mathbf{ar}}{r^3},$$

where **a** is a constant vector. The particle velocity at infinity has an angle α with the vector **a**.

§ 39 Variational principle for the Hamiltonian equations

The Lagrangian equations can be obtained from the principle of least action, as discussed in § 11. The Hamiltonian equations can also be obtained as the Euler equations of some variational problem.

To this end, we introduce a function

$$\Lambda(q, p, \dot{q}, t) = \sum_{i=1}^{s} \dot{q}_i p_i - H(q, p, t), \tag{39.1}$$

in which q_i and p_i are considered as independent variables, so their variations δq_i and δp_i are also considered independent. Note that the function $\Lambda(q, p, \dot{q}, , t)$ does not depend on $\dot{p}$. Let a system of particles at time t_1 be located at the point $A(q_1^{(1)}, \ldots, q_s^{(1)})$, and at time

t_2 be at point $B(q_1^{(2)}, \dots, q_s^{(2)})$. The variations of the generalized coordinates at the points A and B are also assumed to be zero[2]

$$\delta q_i(t_1) = \delta q_i(t_2) = 0, \quad i = 1, 2, \dots, s. \tag{39.2}$$

Then the following statement is true: the motion of the system between these two points occurs according to a law $q_i(t)$, $p_i(t)$, such that when we substitute these functions $q_i(t)$, $p_i(t)$ into the integral

$$\Sigma = \int_{t_1}^{t_2} \Lambda(q, p, \dot{q}, t)\, dt, \tag{39.3}$$

the latter takes an extreme value, i.e. the variation $\delta\Sigma = 0$.

Indeed, the variation of $\delta\Sigma$ is equal to

$$\delta\Sigma = \int_{t_1}^{t_2} \sum_{i=1}^{s} \left[\left(\frac{\partial \Lambda}{\partial q_i} - \frac{d}{dt}\frac{\partial \Lambda}{\partial \dot{q}_i} \right) \delta q_i + \frac{\partial \Lambda}{\partial p_i} \delta p_i \right] dt + \sum_{i=1}^{s} \frac{\partial \Lambda}{\partial \dot{q}_i} \delta q_i \Big|_{t_1}^{t_2}. \tag{39.4}$$

Then, taking into account the boundary conditions (2) and the independence of the coordinate and momentum variations, we obtain the Euler equations for this variational problem

$$\frac{\partial \Lambda}{\partial q_i} - \frac{d}{dt}\frac{\partial \Lambda}{\partial \dot{q}_i} = 0, \quad \frac{\partial \Lambda}{\partial p_i} = 0, \quad i = 1, 2, \dots, s. \tag{39.5}$$

They coincide with the Hamiltonian equations, since

$$\frac{\partial \Lambda}{\partial q_i} = -\frac{\partial H}{\partial q_i}, \quad \frac{\partial \Lambda}{\partial \dot{q}_i} = p_i, \quad \frac{\partial \Lambda}{\partial p_i} = \dot{q}_i - \frac{\partial H}{\partial p_i}. \tag{39.6}$$

Now compare the variational principle $\delta\Sigma = 0$ and the principle of least action $\delta S = 0$ (see § 11). If we define dependency $p_i(t)$ from the equations $p_i = \partial L/\partial \dot{q}_i$, then the function Λ coincides with the Lagrangian function L, and Σ coincides with the action S. However, momenta are independent variables (as well as coordinates) in the variational principle $\delta\Sigma = 0$, and the variations of δq_i and δp_i are also independent. In other words, the class of 'eligible to competition' functions $p_i(t)$ in the principle $\delta\Sigma = 0$ is much wider than in the principle of least action $\delta S = 0$, since for the latter the variations $\delta p_i(t)$ are completely determined by variations of coordinates and their time derivatives.

§ 40 Poisson brackets

Given the Hamiltonian $H(q, p, t)$ of a system, we aim to determine the total time derivative of an arbitrary function of coordinates, momenta, and time, denoted by $f(q, p, t)$.

[2] Variations of the generalized momenta at the time moments t_1 and t_2 can be arbitrary.

To achieve this, we differentiate f with respect to time and express $\dot{q}_i$ and $\dot{p}_i$ from the Hamiltonian equations, yielding the expression:

$$\frac{df(q,p,t)}{dt} = \frac{\partial f}{\partial t} + \sum_{i=1}^{s} \left(\frac{\partial H}{\partial p_i}\frac{\partial f}{\partial q_i} - \frac{\partial H}{\partial q_i}\frac{\partial f}{\partial p_i} \right). \tag{40.1}$$

The bilinear combination of partial derivatives arising in Eq. (1), commonly referred to as the *Poisson bracket*, appears frequently and plays a significant role in the study of problems in Hamiltonian mechanics.

40.1 *Definition and main properties*

Let $f = f(q,p,t)$ and $g = g(q,p,t)$ be arbitrary functions of the generalized coordinates, momenta and time. The *Poisson bracket* $\{f,g\}$ is defined as follows:

$$\{f,g\} \equiv \sum_{i=1}^{s} \left(\frac{\partial f}{\partial p_i}\frac{\partial g}{\partial q_i} - \frac{\partial f}{\partial q_i}\frac{\partial g}{\partial p_i} \right). \tag{40.2}$$

Poisson brackets have a number of properties which can be easily verified using the definition (2):

$$\begin{aligned}
\{f,c\} &= 0, \quad \text{if } c = \text{const}, \\
\{f,g\} &= -\{g,f\}, \\
\{f_1 + f_2, g\} &= \{f_1, g\} + \{f_2, g\}, \\
\{f_1 f_2, g\} &= f_1\{f_2, g\} + \{f_1, g\} f_2, \\
\frac{\partial}{\partial t}\{f,g\} &= \left\{\frac{\partial f}{\partial t}, g\right\} + \left\{f, \frac{\partial g}{\partial t}\right\}, \\
\{p_i, f\} &= \frac{\partial f}{\partial q_i}, \quad \{q_i, f\} = -\frac{\partial f}{\partial p_i}, \\
\{p_i, p_j\} &= \{q_i, q_j\} = 0, \quad \{p_i, q_j\} = \delta_{ij}.
\end{aligned}$$

Using these properties, one can calculate the value of the Poisson brackets without referring to their definition, at least for the polynomial functions f and g.

Now Eq. (1) can be rewritten as

$$\frac{df(q,p,t)}{dt} = \frac{\partial f}{\partial t} + \{H, f\}, \tag{40.3}$$

and, in particular, the Hamiltonian equations themselves (38.8) can also be rewritten in the same way:

$$\dot{q}_i = \{H, q_i\}, \quad \dot{p}_i = \{H, p_i\}. \tag{40.4}$$

Using the Poisson brackets of the form $\{H,f\}$, $\{H,\{H,f\}\}$, ... for $f = q(t)$ or $f = p(t)$, we can obtain a general solution to the Hamiltonian equations, only formally represented as a series in powers of time from the starting moment (see [3], problem 10.24). From this perspective, the Poisson bracket $\{H, f(q,p)\}$ can be said to determine the 'development' (evolution) of the function $f(q(t), p(t))$ in relation to time t.

Example 1

Similarly, the Poisson bracket $\{M_z, f(\mathbf{r}, \mathbf{p})\}$ is involved in the evolution of the function $f(\mathbf{r}, \mathbf{p})$ in relation to the angle of rotation around the z-axis denoted by φ. When a mechanical system as a whole rotates by an infinitely small angle ε around the z-axis, the change δf of an arbitrary function of the coordinates and momenta $f(\mathbf{r}, \mathbf{p})$ in the first order of ε equals[3]

$$\delta f(\mathbf{r}, \mathbf{p}) = \frac{\partial f}{\partial \varphi}\, \varepsilon.$$

On the other hand, according to (10.9), we have $M_z = p_\varphi$, therefore

$$\{M_z, f(\mathbf{r}, \mathbf{p})\} = \frac{\partial M_z}{\partial p_\varphi}\frac{\partial f}{\partial \varphi} = \frac{\partial f}{\partial \varphi}.$$

As a result, we get

$$\delta f(\mathbf{r}, \mathbf{p}) = \{M_z, f(\mathbf{r}, \mathbf{p})\}\, \varepsilon.$$

Consequence 1:

If $f = S(\mathbf{r}^2, \mathbf{p}^2, \mathbf{rp})$ is a scalar function of coordinates and momenta, then it remains unchanged under rotation around the z-axis or any arbitrary axis. Thus,

$$\{\mathbf{M}, S(\mathbf{r}^2, \mathbf{p}^2, \mathbf{rp})\} = 0. \tag{40.5}$$

Consequence 1:

Let $f = V_x$ be the x-component of the vector function of coordinates and momenta, which can be expressed as $\mathbf{V} = \mathbf{r}\, S_1 + \mathbf{p}\, S_2 + [\mathbf{r}, \mathbf{p}]\, S_3$, where $S_i = S_i(\mathbf{r}^2, \mathbf{p}^2, \mathbf{rp})$ are scalar functions. In this case, the change on rotation is given by $\delta V_x = -x V_y$, which means that

$$\{M_z, V_x\} = -V_y.$$

By generalizing this result, we obtain

$$\{M_i, V_j\} = -\sum_{k=1}^{3} e_{ijk} V_k, \tag{40.6}$$

where e_{ijk} is a completely antisymmetric unit tensor of rank three. In particular, we have

$$\{M_i, M_j\} = -\sum_{k} e_{ijk} M_k. \tag{40.7}$$

[3] In a more detailed notation, this equality has the following form:

$$\begin{aligned}\delta f(\mathbf{r}, \mathbf{p}) &= f(x - \varepsilon y,\, y + \varepsilon x,\, z,\, p_x - \varepsilon p_y,\, p_y + \varepsilon p_x,\, p_z) - f(x, y, z, p_x, p_y, p_z) = \\ &= \varepsilon\left(-\frac{\partial f}{\partial x} y + \frac{\partial f}{\partial y} x - \frac{\partial f}{\partial p_x} p_y + \frac{\partial f}{\partial p_y} p_x\right) = \varepsilon \{M_z, f\}.\end{aligned}$$

Example 2

Let us consider the Poisson brackets for the velocity components of a charged particle in a magnetic field and for the coordinates of the centre of the orbit. Suppose that the magnetic field is given by the vector potential $\mathbf{A}(\mathbf{r}, t)$. According to (12.5), the velocity components are given by

$$v_i = \frac{p_i}{m} - \frac{e}{mc}A_i,$$

which leads to

$$\{v_i, v_j\} = -\frac{e}{m^2c}\left(\{p_i, A_j\} + \{A_i, p_j\}\right) =$$

$$= -\frac{e}{m^2c}\left(\frac{\partial A_j}{\partial x_i} - \frac{\partial A_i}{\partial x_j}\right) = -\frac{e}{m^2c}\sum_{k=1}^{3} e_{ijk}B_k,$$

where $\mathbf{B} = \text{rot}\mathbf{A}$ is the magnetic field.

Assuming that the magnetic field is uniform and constant and restricting the particle motion to a plane perpendicular to the magnetic field, we can choose the z-axis to be along the $\mathbf{B}$ direction. Then, the only non-zero Poisson bracket is given by

$$\{v_x, v_y\} = -\frac{eB_z}{m^2c} = -\frac{\omega}{m}, \quad \omega = \frac{eB}{mc}. \tag{40.8}$$

The particle moves along a circle centred at the point (x_0, y_0) with angular velocity $\boldsymbol{\omega} = -(e/mc)\mathbf{B}$ (see (38.16)–(38.20)), while

$$x_0 = x - \frac{v_y}{\omega}, \quad y_0 = y + \frac{v_x}{\omega}. \tag{40.9}$$

Taking into account that $\{mv_i, x_j\} = \delta_{ij}$, we find the Poisson bracket for the coordinates of the orbit centre:

$$\{x_0, y_0\} = \frac{1}{m\omega}. \tag{40.10}$$

40.2 *Jacobi identity and Poisson's theorem*

For any three functions of the coordinates and momenta $f(q,p)$, $g(q,p)$, $h(q,p)$ the following relations are valid

$$\{f, \{g, h\}\} + \{g, \{h, f\}\} + \{h, \{f, g\}\} = 0, \tag{40.11}$$

which are called the *Jacobi identity*.

To prove this identity, we use a formal trick, noting that the Jacobi identity only involves derivatives with respect to q and p, and not time. We assume $h(q,p)$ to be the 'Hamiltonian' of an imaginary mechanical system, which means that the development of q_i and p_i, varying with the imaginary time parameter τ (not to be confused with real time t), is determined by the 'canonical equations' generated by the function $h(q,p)$:

$$\frac{dq_i}{d\tau} = \frac{\partial h}{\partial p_i}, \quad \frac{dp_i}{d\tau} = -\frac{\partial h}{\partial q_i}. \tag{40.12}$$

From these equations, q_i, p_i and $g(q(\tau), p(\tau))$ can be found as functions of τ. Besides, in accordance with (3), we can exploit equations

$$\frac{df}{d\tau} = \{h, f\}, \quad \frac{dg}{d\tau} = \{h, g\}, \quad \frac{d\{f,g\}}{d\tau} = \{h, \{f, g\}\}. \tag{40.13}$$

Using the properties of the Poisson brackets, we find

$$\frac{d\{f,g\}}{d\tau} = \left\{\frac{df}{d\tau}, g\right\} + \left\{f, \frac{dg}{d\tau}\right\}. \tag{40.14}$$

Substituting formulae (13) into this relation, we obtain the Jacobi identity (11).

The Jacobi identity remains valid even if the functions f, g, and h also have explicit time dependence such that $f = f(q,p,t)$, $g = g(q,p,t)$, and $h = h(q,p,t)$. In this case, time t is a parameter that is not relevant to the calculation of partial derivatives with respect to p and q.

A more direct (but also more cumbersome) way to prove the Jacobi identity is given, for example, in [1], § 41.

Poisson's theorem. *If a mechanical system has two integrals of motion $f(q,p,t)$ and $g(q,p,t)$, then the Poisson bracket $h = \{f, g\}$ is also the integral of motion for this system.*

Indeed, from the conditions

$$\frac{df}{dt} = \frac{\partial f}{\partial t} + \{H, f\} = 0,$$
$$\frac{dg}{dt} = \frac{\partial g}{\partial t} + \{H, g\} = 0,$$

the Jacobi identity (11) and the properties of the Poisson brackets we immediately obtain the equality

$$\frac{dh}{dt} = \frac{\partial h}{\partial t} + \{H, h\} = 0.$$

As an example, consider the integrals of motion for a particle in the field of an isotropic three-dimensional oscillator $U(r) = kr^2/2$. It is known (see § 6) that the angular momentum

$$\boldsymbol{M} = (0,\ 0,\ M), \quad M = xp_y - yp_x$$

is conserved in such a field, as are the energies of oscillations along the x- and y-axes:

$$E_x = \frac{p_x^2}{2m} + \frac{kx^2}{2}, \quad E_y = \frac{p_y^2}{2m} + \frac{ky^2}{2}.$$

According to Poisson's theorem, the quantity

$$N = \{E_x, M\} = \frac{p_x p_y}{m} + kxy$$

is also an integral of motion, as well as the quantity $\{E_y, M\} = -N$. *That has already been noted earlier (see Eq. (6.7)). Repeated application of the Poisson's theorem does not lead to new integrals of motion, thus*

$$\{E_x, N\} = -\{E_y, N\} = -\frac{k}{m} M.$$

Problem

40.1. Evaluate the Poisson brackets:
a) $\{M_i, x_j\}$, $\{M_i, p_j\}$, $\{M_i, \mathbf{aM}\}$;
b) $\{\mathbf{ap}, \mathbf{br}\}$, $\{\mathbf{aM}, \mathbf{br}\}$, $\{\mathbf{aM}, \mathbf{bM}\}$;
c) $\{\mathbf{M}, \mathbf{rp}\}$, $\{\mathbf{p}, r^n\}$, $\{\mathbf{p}, ,(\mathbf{ar})^2\}$.
Here x_i, p_i, M_i are the Cartesian components of the vectors, while **a** and **b** are the constant vectors.

§ 41 Dynamic symmetry of the Kepler problem

Noether's theorem, as discussed in § 17.3, provides a means of determining the integral of motion that corresponds to the known symmetry of a problem. An arbitrary central field, expressed as $U(r)$, exhibits a spherical symmetry in three-dimensional space and conserves the three angular momentum components, M_x, M_y, and M_z, relating to rotations in the yz-, zx-, and xy-planes. In the case of the Kepler problem where a particle moves in a field of the form $U(r) = -\alpha/r$, an additional integral of motion exists. This is the Laplace vector, defined as follows:

$$\mathbf{A} = [\mathbf{v}, \mathbf{M}] - \frac{\alpha\mathbf{r}}{r} = \mathbf{r}(\mathbf{vp}) - \mathbf{p}(\mathbf{vr}) - \frac{\alpha\mathbf{r}}{r}.$$

Now, consider a finite motion of a particle with total energy $H = E < 0$. This specific case shows that the vectors **A** and **M** can be used to construct two *independent* vectors:

$$\mathbf{J}_{1,2} = \frac{1}{2}\left(\mathbf{M} \pm \sqrt{\frac{m}{-2E}}\,\mathbf{A}\right), \tag{41.1}$$

Each of these vectors possesses a certain similarity to the vector **M** as well as similar Poisson brackets.

To demonstrate this, let us represent the Laplace vector **A** as the sum:

$$\mathbf{A} = 2E\mathbf{r} + \mathbf{C}, \quad \mathbf{C} = -\mathbf{p}(\mathbf{vr}) + \frac{\alpha\mathbf{r}}{r}.$$

Poisson brackets, as outlined in § 40.1, allow us to determine that:

$$\{M_i, M_j\} = -\sum_k e_{ijk} M_k, \quad \{M_i, A_j\} = -\sum_k e_{ijk} A_k, \tag{41.2}$$

and that:

$$\{x_i, C_j\} + \{C_i, x_j\} = (x_i p_j - x_j p_i)/m = \sum_k e_{ijk} M_k/m, \quad \{C_i, C_j\} = 0,$$

thereby indicating:

$$\{A_i, A_j\} = \frac{2E}{m} \sum_k e_{ijk} M_k. \tag{41.3}$$

Taking into account relations (2)–(3), we obtain the following properties of $\mathbf{J}_{1,2}$

$$\{H, \mathbf{J}_{1,2}\} = 0, \quad \{J_{1i}, J_{2j}\} = 0, \tag{41.4}$$

$$\{J_{1i}, J_{1j}\} = -\sum_k \epsilon_{ijk} J_{1k}, \quad \{J_{2i}, J_{2j}\} = -\sum_k \epsilon_{ijk} J_{2k}. \tag{41.5}$$

Therefore, the vectors $\mathbf{J}_1$ and $\mathbf{J}_2$ represent independent integrals of motion. Moreover, each vector exhibits the same Poisson brackets for their component as the standard angular momentum. The two 'angular momenta' refer to the symmetry of rotations in six planes, which corresponds to the spherical symmetry of a four-dimensional space. These properties of the Kepler problem are closely related to the so-called 'hidden or dynamic symmetry' of the hydrogen atom, as described in [15], Chapter I, § 5.

In conclusion, note that the Hamiltonian function H can be expressed in terms of $\mathbf{J}_1^2 = \mathbf{J}_2^2 = -m\alpha^2/(8E)$, such that:

$$H = -\frac{m\alpha^2}{4(\mathbf{J}_1^2 + \mathbf{J}_2^2)}. \tag{41.6}$$

§ 42 Classical model of EPR and NMR

Electron paramagnetic resonance (EPR) was discovered by E.K. Zavoisky in 1944, and two years later, in 1946, F. Bloch and E. Purcell discovered nuclear magnetic resonance (NMR). Both phenomena involve the resonant absorption of electromagnetic waves with wavelengths ranging from 0.01 to 10 cm for EPR and 10 to 100 m for NMR, due to the paramagnetism of electrons or nuclei. EPR and NMR have been widely used for fundamental research in solid state physics, chemistry, and biology. Additionally, they have numerous applications in engineering and medicine (see [16], Ch. IX).

In this section, we consider a simple classical model of EPR and NMR. The model is in the form of a uniform magnetized rotating ball with the moment of inertia I and a magnetic moment equal to $g\mathbf{M}$, where g is the gyromagnetic ratio, and $\mathbf{M}$ is the angular momentum of the ball. The Hamilton function of a ball in a homogeneous magnetic field $\mathbf{B}(t)$ is given by:

$$H = \frac{M^2}{2I} - g\mathbf{M}\mathbf{B}(t). \tag{42.1}$$

42.1 *Equations of motion of the vector* $\mathbf{M}(t)$

The equation of motion of the angular momentum $\mathbf{M}(t)$ is convenient to write using the Poisson brackets

$$\frac{d\mathbf{M}}{dt} = \{H, \mathbf{M}\} = \frac{\{M^2, \mathbf{M}\}}{2I} - g\sum_i \{M_i, \mathbf{M}\}\, B_i(t).$$

When evaluating the z-component of this equation, we need to compute the following Poisson brackets (see § 40.1)

$$\{M^2, M_z\} = 0, \quad \{M_x, M_z\} = M_y, \quad \{M_y, M_z\} = -M_x, \quad \{M_z, M_z\} = 0,$$

which leads to the equation

$$\frac{dM_z}{dt} = -g\,(M_y B_x - M_x B_y) = g[\mathbf{M}, \mathbf{B}(t)]_z$$

or

$$\frac{d\mathbf{M}}{dt} = g[\mathbf{M}, \mathbf{B}(t)]. \tag{42.2}$$

It follows that the vector $\mathbf{M}$ rotates with the angular velocity

$$\boldsymbol{\omega}_B(t) = -g\mathbf{B}(t). \tag{42.3}$$

In particular, if the magnetic field is constant and directed along the z-axis,

$$\mathbf{B} = (0,\, 0,\, B_0), \tag{42.4}$$

then the vector $\mathbf{M}(t)$ precesses around the direction $\mathbf{B}$:

$$\begin{aligned} M_x &= M_x(0)\cos(gB_0t) + M_y(0)\sin(gB_0t), \\ M_y &= -M_x(0)\sin(gB_0t) + M_y(0)\cos(gB_0t), \\ M_z &= M_z(0). \end{aligned}$$

42.2 *Motion of the vector* $\mathbf{M}(t)$ *in a rotating magnetic field*

In this section, we examine the motion of the vector $\mathbf{M}(t)$ in the presence of a magnetic field with a large constant component B_0 along the z-axis and a small component B_1 rotating in the xy-plane with the constant angular velocity Ω:

$$\mathbf{B}(t) = (B_1 \cos\Omega t,\, B_1 \sin\Omega t,\, B_0). \tag{42.5}$$

Let in the initial moment of time the vector $\mathbf{M}(t)$ be directed along the z-axis

$$\mathbf{M}(0) = (0,\, 0,\, M_0). \tag{42.6}$$

Thereafter, the motion of $\mathbf{M}(t)$ represents a rotation with an angular velocity $\boldsymbol{\omega}_B = -g\mathbf{B}(t)$, which, in turn, rotates about the z-axis with an angular velocity Ω.

This complex movement can be conveniently described in the rotating reference frame K'. In K', the vector $\mathbf{B}$ is stationary and its components are equal to the components of the vector (5) in the initial moment of time:

$$B'_x = B_1, \ \ B'_y = 0, \ \ B'_z = B_0. \tag{42.7}$$

The vector $\mathbf{M}$ in the K' frame rotates with a constant angular velocity[4]

$$\omega'_B = -g\mathbf{B}' - \Omega, \tag{42.8}$$

whose components are equal to

$$(\omega'_B)_x = -gB_1, \quad (\omega'_B)_y = 0, \quad (\omega'_B)_z = -gB_1 - \Omega \equiv \epsilon. \tag{42.9}$$

Of particular interest is the resonant case $\varepsilon = 0$ when the vector $\mathbf{M}$ in the frame K' rotates around the x'-axis with the angular velocity $(-gB_1)$,

$$M'_x = 0, \ \ M'_y = M_0 \sin(gB_1 t), \ \ M'_z = M_0 \cos(gB_1 t),$$

that is, when after time $\pi/(gB_1)$ the vector $\mathbf{M}$ will be directed along $(-z)$-axis (Fig. 68). This is called the 'total flip' (in quantum theory, this corresponds to the energy absorption associated with the transition to the nearest upper level).

Generally, for this initial condition, the components $\mathbf{M}$ in a rotating system are equal to

$$\begin{aligned} M'_x &= -\frac{\epsilon}{\lambda} a M_0 (1 - \cos \lambda t), \\ M'_y &= a M_0 \sin \lambda t, \\ M'_z &= \left(\frac{\epsilon^2}{\lambda^2} + a^2 \cos \lambda t \right) M_0, \end{aligned}$$

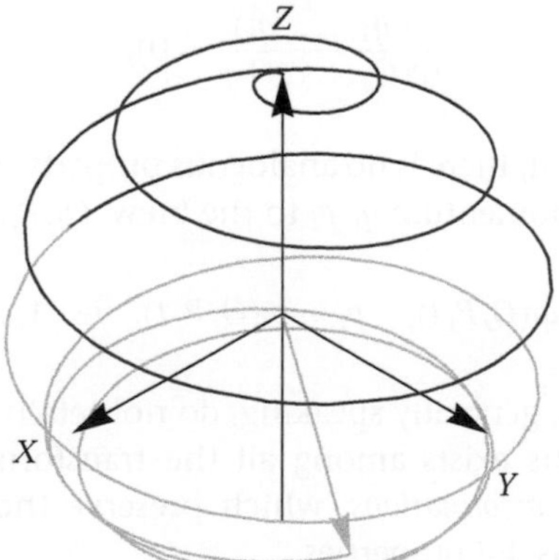

Figure 68 Motion of the angular momentum

[4] This is easy to see from the fact that the Hamiltonian in the rotating frame H' is related to the Hamiltonian (1) by the relation $H' = H - \mathbf{\Omega M}$ (see problem 2 to § 40 from [1].

where

$$\lambda = \sqrt{\epsilon^2 + (gB_1)^2}, \quad a = \frac{gB_1}{\lambda}.$$

In a stationary system, the components of **M** are described by:

$$M_x = M'_x \cos\Omega t - M'_y \sin\Omega t,$$
$$M_y = M'_x \sin\Omega t + M'_y \cos\Omega t,$$
$$M_z = M'_z.$$

For $B_1 \ll B_0$, the dependence of the amplitudes $M_{x,y}$ on Ω has a resonant character: generally speaking, these amplitudes are small $\sim M_0 B_1/B_0$, but for $|\epsilon| = |\Omega_B + gB_1| \lesssim gB_1$ they sharply increase reaching values $\sim M_0$. In particular, when $\Omega_B = -gB_0$, we have

$$M_x = M_0 \sin(gB_1 t)\sin(gB_0 t),$$
$$M_y = M_0 \sin(gB_1 t)\cos(gB_0 t),$$
$$M_z = M_0 \cos(gB_1 t).$$

§ 43 Canonical transformations

One of the attractive features of the Lagrangian formalism is the covariance of the Lagrangian equations under transformations of the generalized coordinates, i.e. transition from one set of the generalized coordinates q_i to any other Q_i:

$$q_i = q_i(Q, t), \quad i = 1, \ldots, s$$

(of course, this requires that the transformation to be non-degenerate:

$$\frac{\partial(q_1, \ldots, q_s)}{\partial(Q_1, \ldots, Q_s)} \neq 0).$$

In the Hamiltonian approach, there is no analogous property: for arbitrary transition from the 'old' coordinates and momentum q_i, p_i to the 'new' Q_i, P_i, i.e. when replacing

$$q_i = q_i(Q, P, t), \quad p_i = p_i(Q, P, t), \quad i = 1, \ldots, s, \tag{43.1}$$

the Hamiltonian equations, generally speaking, do not retain their form. However, a special class of transformations exists among all the transformations of the given form, known as the *canonical transformations*, which preserve the form of the equations of motion and possess other useful properties.

43.1 *Definition of the canonical transformation. Generating function*

Before defining the canonical transformation, two preliminary remarks should be made. Firstly, transformation of the form (1) involves only coordinates and momenta, with no influence on time t, which is a parameter of the transformation.[5] Secondly, while the definition of the canonical transformation is not exactly obvious, it provides a quick access to a range of useful properties.

So, here is the definition: *the canonical transformation* is the transformation of the form (1) such that

$$\sum_{i=1}^{s} p_i dq_i = \sum_{i=1}^{s} P_i dQ_i + dF(q, Q, \breve{t}), \tag{43.2}$$

where $dF(q, Q, \breve{t})$ is the total differential of the function $F(q, Q, t)$ at a fixed time, that is,

$$dF(q, Q, \breve{t}) \equiv dF(q, Q, t)_{t=\mathrm{const}} = dF(q, Q, t) - \frac{\partial F}{\partial t} dt. \tag{43.3}$$

The function $F(q, Q, t)$ is called the *generating function* of the given canonical transformation. From definition (2), it immediately follows

$$p_i = \frac{\partial F}{\partial q_i}, \quad P_i = -\frac{\partial F}{\partial Q_i}. \tag{43.4}$$

Upon resolving these algebraic relations, explicit equations for connection of the old and new variables can be obtained using the known generating function $F(q, Q, t)$.

Now, the covariance of the Hamiltonian equations with respect to canonical transformations can be proved. A quick reminder is due: the Hamiltonian equations

$$\dot{q}_i = \frac{\partial H}{\partial p_i}, \quad \dot{p}_i = -\frac{\partial H}{\partial q_i}$$

are necessitated by the variational principle (see § 39)

$$\delta\Sigma = \int_{t_1}^{t_2} \sum_i \left[\left(\dot{q}_i - \frac{\partial H}{\partial p_i} \right) \delta p_i - \left(\dot{p}_i + \frac{\partial H}{\partial q_i} \right) \delta q_i \right] dt + \sum_i p_i \, \delta q_i \Big|_{t_1}^{t_2} = 0 \tag{43.5}$$

when we take into account the boundary conditions (39.2) and the independence of the variations of δq_i and δp_i. By virtue of (2), (3), the functional Σ is equal to

$$\Sigma = \int_{t_1}^{t_2} \left(\sum_i P_i \dot{Q}_i - H'(P, Q, t) \right) dt + \int_{t_1}^{t_2} dF(q, Q, t),$$

where

$$H'(Q, P, t) = H(q, p, t) + \frac{\partial F(q, Q, t)}{\partial t}. \tag{43.6}$$

[5] It is possible to generalize the canonical transformations so that they will also allow the time transformation – see Supplement D.

Therefore, the variational principle (5) can be rewritten as

$$\delta\Sigma = \int_{t_1}^{t_2} \sum_i \left[\left(\dot{Q}_i - \frac{\partial H'}{\partial P_i} \right) \delta P_i - \left(\dot{P}_i + \frac{\partial H'}{\partial Q_i} \right) \delta Q_i \right] dt +$$

$$+ \sum_i \left(P_i \delta Q_i + \frac{\partial F}{\partial q_i} \delta q_i + \frac{\partial F}{\partial Q_i} \delta Q_i \right) \Big|_{t_1}^{t_2} = 0.$$

Due to relations (4) and boundary conditions (39.2), the non-integral terms disappear. Taking into account the independence of the variations δQ_i and δP_i, we obtain the equations of motion for new variables

$$\dot{Q}_i = \frac{\partial H'}{\partial P_i}, \quad \dot{P}_i = -\frac{\partial H'}{\partial Q_i} \tag{43.7}$$

with the new Hamiltonian $H'(Q, P, t)$ defined by Eq. (6). They also have the form of the Hamiltonian equations.

Thus, the generating function $F(q, Q, t)$ defines the canonical transformation using formulae (4). These formulae and relation (6) can be represented in the following compact form:

$$dF(q, Q, t) = \sum_i (p_i dq_i - P_i dQ_i) + (H' - H)dt. \tag{43.8}$$

Example

Imagine a one-dimensional system with the Hamiltonian $H(q, p, t)$. To give a non-trivial example of canonical transformations, we can consider a linear transformation of coordinate and momentum obtained from the generating function $F(q, Q, t)$ which is the quadratic form of the canonical variables:

$$F(q, Q, t) = \frac{1}{2} \left(a q^2 + 2b qQ + c Q^2 \right) \tag{43.9}$$

(here the quantities a, b and c are either constants or functions of time). Using Eqs. (4), we find the relations

$$p = \frac{\partial F}{\partial q} = a q + b Q, \quad P = -\frac{\partial F}{\partial Q} = -b q - c Q.$$

Solving them, we obtain the linear equations for connections between old and new variables:

$$Q = \alpha q + \beta p, \quad P = \gamma q + \delta p, \tag{43.10}$$

in which the coefficients are equal to

$$\alpha = -\frac{a}{b}, \quad \beta = \frac{1}{b}, \quad \gamma = \frac{ac - b^2}{b}, \quad \delta = -\frac{c}{b}. \tag{43.11}$$

Note that these coefficients are not arbitrary, namely, the Jacobian of transformation (10) is equal to unity[6]

$$\frac{\partial(Q,P)}{\partial(q,p)} = \alpha\delta - \beta\gamma = 1. \tag{43.12}$$

The new Hamiltonian defined by Eq. (6) has the form

$$H'(Q,P,t) = H(q,p,t) + \frac{1}{2}\left(\dot{a}\,q^2 + 2\dot{b}\,qQ + \dot{c}\,Q^2\right), \tag{43.13}$$

moreover, on the right side of this equality, it is necessary to replace the old variables q and p by new variables Q and P according to relations (10).

43.2 *Other generating functions*

In addition to the generating function $F(q,Q,t)$, other generating functions can be introduced using variables q,P, or Q,p, or p,P. For example, substituting into (8) the relation

$$P_i dQ_i = d(P_i Q_i) - Q_i dP_i$$

and introducing the generating function

$$\Phi(q,P,t) = F + \sum_{i=1}^{s} Q_i P_i, \tag{43.14}$$

we get

$$d\Phi(q,P,t) = \sum_{i=1}^{s} (p_i dq_i + Q_i dP_i) + (H' - H)\,dt. \tag{43.15}$$

Hence, for the generating function $\Phi(q,P,t)$, formulae similar to (4), (6) are

$$p_i = \frac{\partial\Phi}{\partial q_i}, \quad Q_i = \frac{\partial\Phi}{\partial P_i}, \quad H' = H + \frac{\partial\Phi}{\partial t}. \tag{43.16}$$

Here are some simple examples. The generating function

$$F(q,Q,t) = \sum_{i=1}^{s} q_i Q_i$$

exchanges coordinates and momenta up to sign:

$$Q_i = p_i, \; P_i = -q_i.$$

For this reason, in the Hamiltonian method, the terms 'coordinates' and 'momenta' become conditional and the variables q, p are often referred to as the *canonically conjugate quantities*.

[6] Thus, condition (12) is necessary in order that the linear transformation (10) becomes the canonical one. Below we show that (12) is also a sufficient condition (see example 1 in § 45).

The generating function

$$\Phi(q, P, t) = \sum_{i=1}^{s} q_i P_i$$

defines the identical transformation:

$$Q_i = q_i, \ P_i = p_i.$$

The generating function of a transformation close to the identity can often be represented as

$$\Phi(q, P, t) = \sum_i q_i P_i + \varepsilon W(q, P),$$

where the parameter ε is determined as the smallness of the amendments.

An important example

As an interesting example, we consider the generating function

$$\Phi(q, P, t) = qP + \delta\tau H(q, P, t) \tag{43.17}$$

where $\varepsilon \equiv \delta\tau$ is an infinitesimal parameter, and $H(q, p, t)$ is the Hamiltonian of the system. In this case, the equations for connections of old and new variables (16) have the form

$$p = \frac{\partial \Phi}{\partial q} = P + \delta\tau \frac{\partial H(P, q, t)}{\partial q},$$
$$Q = \frac{\partial \Phi}{\partial P} = q + \delta\tau \frac{\partial H(P, q, t)}{\partial P}.$$

Using the smallness of the $\delta\tau$ parameter, we can rewrite these equations in the form

$$P(t) = p - \delta\tau \frac{\partial H(q, p, t)}{\partial q} = p(t) + \dot{p}\delta\tau = p(t + \delta\tau),$$
$$Q(t) = q + \delta\tau \frac{\partial H(q, p, t)}{\partial p} = q(t) + \dot{q}\delta\tau = q(t + \delta\tau).$$

Thus, the considered transformation

$$Q(t) = q(t + \delta\tau), \ \ P(t) = p(t + \delta\tau) \tag{43.18}$$

is a time shift by $\delta\tau$ (cf. [1], § 45). The new Hamiltonian

$$\begin{aligned} H'(Q, P, t) &= H(q, p, t) + \frac{\partial \Phi}{\partial t} = H(q, p, t) + \delta\tau \frac{\partial H(q, P, t)}{\partial t} \approx \\ &\approx H(q, p, t) + \delta\tau \frac{dH(q, p, t)}{dt} = H(q, p, t + \delta\tau) \end{aligned}$$

also corresponds to a time shift of $\delta\tau$. From here it follows that, making successive small shifts in time, one can obtain a finite time shift and show that transformation

$$Q(t) = q(t+\tau), \quad P(t) = p(t+\tau) \tag{43.19}$$

is also canonical. The generating function for this transformation will be specified in § 46.2.

To conclude this section, it is worth mentioning that the canonical transformations look especially simple in formalism of the *external differential forms*. This issue is considered in Supplement E. In particular, in this formalism it is easy to prove the following two important properties of canonical transformations: (i) the Poisson brackets are invariant under canonical transformations, and (ii) the Jacobian of the canonical transformation (1) is equal to unity (this means that under canonical transformations the phase volume of system is conserved). In the standard approach, the proof of these properties is significantly more cumbersome (see § 44 and § 47 below).

Problems

43.1. Find out the meaning of canonical transformations given by generating functions:
a) $\Phi(\mathbf{r}, \mathbf{P}) = \mathbf{rP} + \delta\mathbf{aP}$;
b) $\Phi(\mathbf{r}, \mathbf{P}) = \mathbf{rP} + \delta\boldsymbol{\varphi}[\mathbf{r}, \mathbf{P}]$.
Here $\mathbf{r}$ are the Cartesian coordinates, and $\delta\mathbf{a}$ and $\delta\boldsymbol{\varphi}$ are the infinitesimal parameters.

43.2. Show that the canonical transformation given by generating function

$$\Phi(x, y, P_x, P_y) = xP_x + yP_y + \varepsilon(xy + P_xP_y),$$

where $\varepsilon \to 0$, represents rotation in the phase space.

§ 44 Canonical transformations and Poisson brackets

There is a close relationship between the canonical transformations and the Poisson brackets. In this section, two important properties of the Poisson brackets will be discussed. First, we will show that the Poisson brackets are invariant under the canonical transformations. Secondly, a simple and constructive method of checking whether the given transformation is canonical will be presented. This method makes use of the so-called *fundamental Poisson brackets*.

44.1 *Invariance of the Poisson brackets with respect to the canonical transformations*

Let us prove that *the Poisson brackets are invariant under the canonical transformations*. It is convenient to use the same formal trick as in the proof of the Jacobi identity in § 40.2. Let f and h be some functions of q, p, and t. In the canonical transformations, as was mentioned earlier, time is a parameter, so we will further consider $t = \text{const}$. Imagine some mechanical system with $h(q, p, t)$ as the 'Hamiltonian'. Similar to § 40.2, this assumption means that the development of q_i and p_i depending on imaginary 'time' τ (no way associated with the real time t) is determined by the 'canonical equations' (40.12), from which one can find q_i and p_i as functions of τ, and hence also $f(q(\tau), p(\tau), t)$ as the function of τ. Moreover, in accordance with (40.3), we have

$$\frac{df}{d\tau} = \{h, f\}_{p,q} \equiv \sum_i \left(\frac{\partial h}{\partial p_i}\frac{\partial f}{\partial q_i} - \frac{\partial h}{\partial q_i}\frac{\partial f}{\partial p_i} \right). \tag{42.1}$$

The 'time' τ is not explicitly included in the canonical transformation $q, p \to Q, P$ of the form (43.1). From this follows the new 'Hamiltonian'

$$h'(Q, P, t) = h(q(Q, P, t), p(Q, P, t), t). \tag{44.2}$$

Therefore, the derivative with respect to τ of

$$f'(Q, P, t) = f(q(Q, P, t), p(Q, P, t), t), \tag{44.3}$$

is given by the expression

$$\frac{df'}{d\tau} = \{h', f'\}_{P,Q} \equiv \sum_i \left(\frac{\partial h'}{\partial P_i}\frac{\partial f'}{\partial Q_i} - \frac{\partial h'}{\partial Q_i}\frac{\partial f'}{\partial P_i} \right). \tag{44.4}$$

Taking into account (2) and (3), the left parts of Eqs. (1) and (4) coincide, as, in turn, do the right parts, which leads to the desired equality:

$$\{h, f\}_{p,q} = \{h, f\}_{P,Q}. \tag{44.5}$$

44.2 *Necessary and sufficient criterion that the transformation is canonical*

The definition of canonical transformations, as given in § 43.1, lacks visual clarity and is often ineffective in checking whether a given transformation is canonical. Hence, it is crucial to develop a simple and constructive method to verify the canonicity of a transformation. It has been proven that the transformation $q, p \to Q, P$ of the form (43.1) is canonical if and only if the following equalities hold for the so-called *fundamental Poisson brackets*:

$$\{Q_i, Q_j\}_{p,q} = 0, \quad \{P_i, P_j\}_{p,q} = 0, \quad \{P_i, Q_j\}_{p,q} = \delta_{ij}. \tag{44.6}$$

A rather cumbersome proof of this assertion can be found in [2]. Here we solely establish the necessity of this criterion. If a given transformation is canonical, the relations (6) should automatically follow from the equalities (5). For example:

$$\{Q_i, Q_j\}_{p,q} = \{Q_i, Q_j\}_{P,Q} = \sum_k \left(\frac{\partial Q_i}{\partial P_k}\frac{\partial Q_j}{\partial Q_k} - \frac{\partial Q_i}{\partial Q_k}\frac{\partial Q_j}{\partial P_k} \right) = 0, \tag{44.7}$$

since $\partial Q_l/\partial P_k = 0$ due to independence of canonical coordinates and momenta.

§ 45 Examples of canonical transformations

Example 1

Using conditions (44.5), it is easy to show that the linear transformation

$$Q = \alpha x + \beta p, \quad P = \gamma x + \delta p, \tag{45.1}$$

where α, β, γ, δ are complex numbers or functions of time, is canonical if

$$\{P,\, Q\}_{p,x} = \alpha\delta - \beta\gamma = 1.$$

Example 2

A special case of transformation (1) is a rotation by angle φ in the x, $p/(m\omega)$ plane (Fig. 69). In the case of a harmonic oscillator, such a rotation (clockwise!) with $\varphi = \omega t$,

$$x = A\cos(\omega t + \varphi_0), \quad p = -m\omega A \sin(\omega t + \varphi_0), \tag{45.2}$$

leads to new canonical variables

$$Q = A\cos\varphi_0, \quad P = -m\omega A \sin\varphi_0,$$

independent of time. As a result, the new Hamiltonian is zero:

$$H' = 0.$$

These canonical variables, while not interesting in themselves, are nonetheless very useful as a first step in constructing perturbation theories when the initial Hamiltonian differs from the Hamiltonian of the harmonic oscillator by small amendments (see, for example, [3], problem 11.30).

Example 3

In the quantum theory of the harmonic oscillator, the following combinations of coordinate and momentum play an important role:

$$a = \frac{m\omega x + ip}{\sqrt{2m\omega}}, \quad a^* = \frac{m\omega x - ip}{\sqrt{2m\omega}}. \tag{43.3}$$

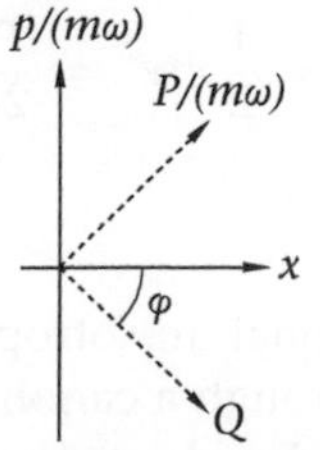

Figure 69 Rotation on the angle φ clockwise in the x, $p/(m\omega)$ plane

(they correspond to the operators of annihilation and creation of quanta). It is easy to check that

$$Q = a, \quad P = ia^* \tag{43.4}$$

are canonical variables that can be expressed in terms of time-dependent variables as follows:

$$Q = a = \sqrt{\frac{m\omega}{2}} A\, e^{-i(\omega t + \varphi_0)}, \quad P = ia^* = i\sqrt{\frac{m\omega}{2}} A\, e^{+i(\omega t + \varphi_0)}, \tag{45.3a}$$

where A and φ_0 are constants. The corresponding Hamiltonian takes on a simple form:

$$H' = \omega a^* a = -i\omega QP. \tag{45.3b}$$

One can see that the canonical variables are also given by the quantities

$$ae^{i\omega t} \text{ and } ia^* e^{-i\omega t},$$

which for the harmonic oscillator do not depend on time and lead to a new Hamiltonian of $H' = 0$. The use of such variables is illustrated in a number of problems on non-linear oscillations in [3], such as problems 11.26 and 11.28.

Example 4

For a particle in a uniform magnetic field directed along the z-axis, $\mathbf{B} = (0, 0, B)$, the following canonical transformation from the initial variables x, y, p_x, and p_y can be used:

$$\begin{aligned} P_X &= p_x = mv_x, \quad X = x - \frac{p_y}{m\omega} = -\frac{v_y}{\omega}, \\ P_Y &= p_y = m\omega x_0, \quad Y = y - \frac{p_x}{m\omega} = y_0, \end{aligned}$$

where $\omega = eB/(mc)$ (see example 1 in § 38.2). It is easy to verify that this transformation is canonical using the fundamental Poisson brackets.

If the magnetic field is constant, the new Hamiltonian can be expressed in terms of the oscillator Hamiltonian as:

$$H(X, Y, P_X, P_Y,) = \frac{1}{2} m\mathbf{v}^2 = \frac{P_X^2}{2m} + \frac{1}{2} m\omega^2 X^2.$$

Example 5

The Hamiltonian of a three-dimensional anisotropic harmonic oscillator in a uniform magnetic field can be transformed through a canonical transformation involving a rotation by the same angle in the $x, p_y/(m\omega)$ and $y, p_x/(m\omega)$ planes. The resulting Hamiltonian can be expressed as a sum of squares (see [3], problems 11.8–11.10).

Problem

45.1. Consider small oscillations of an anharmonic oscillator whose Hamiltonian is

$$H = \frac{p^2}{2m} + \frac{1}{2} m\omega^2 x^2 + \alpha x^3,$$

and whose oscillation amplitudes A satisfy the condition $|\alpha A^3| \ll m\omega^2 A^2$.

Perform a near-identity canonical transformation using the generating function

$$\Phi(x, P, t) = xP + ax^3 + bx^2 P + cxP^2 + dP^3.$$

It can be shown that $a = c = 0$, and that suitable choices of parameters b and d result in a new Hamiltonian, which is the Hamiltonian of a harmonic oscillator up to third-order terms in the new variables Q and P inclusive. Find the expression for $x(t)$.

§ 46 Action along the true trajectory as a function of initial and final coordinates and time

Consider a particle with a generalized coordinate $q^{(1)}$ at the initial time t_1 (referred to as *point 1* in Fig. 70). Different initial particle velocities correspond to various laws of motion (see curves $a, b, c, \ldots$). Let *point 2* in Fig. 70 with coordinates q and time t correspond to some curve a. The integral

$$S_{12} = \int_1^2 L(q, \dot{q}, t)\, dt, \tag{46.1a}$$

taken along the curve a from point 1 to point 2 is the function of the endpoint 2 (for fixed value t_1 and $q^{(1)}$). This is to say,

$$S_{12} = S(q, t). \tag{46.1b}$$

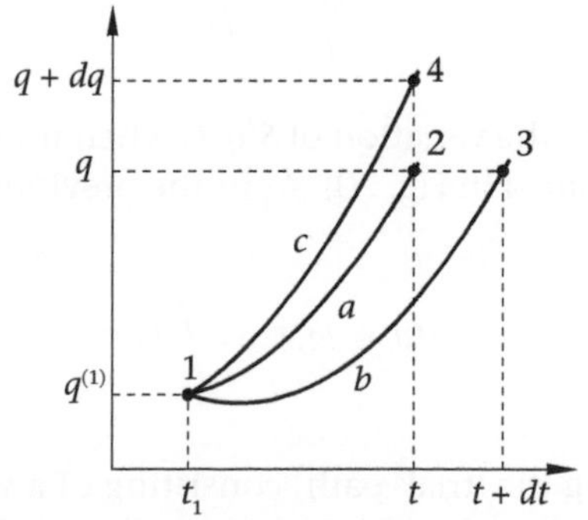

Figure 70 On defining an action as a function of coordinates and time and its partial derivatives

Let us analyse this function in detail. Unlike the integral of action considered in variational principles, the true law of motion is used in Eq. (1). This function happens to offer another way for solving problems of mechanics which differs from the Newtonian, Lagrangian, and Hamiltonian approaches. This approach allows for an analogy between problems of mechanics and optics (see below § 48.3). Moreover, the function $S(q,t)$ is useful in quantum mechanics where, in the semiclassical approximation, the quantity $S(q,t)/\hbar$ coincides with the phase of the wave function; here $\hbar$ is Planck's constant.

46.1 *Properties of* $S(q,t)$

In this section, we will find the partial derivatives of the function $S(q,t)$. Let us start with $\partial S/\partial t$. Keeping in mind the principle of least action, let us consider two different paths leading from *point 1* to *point 3* with coordinate q and time $t+dt$: one path is along the curve of the true motion b (with contribution $S(q,t+dt)$), the second path consists of a segment $1\to 2$ along curve a (with contribution $S(q,t)$), and a small segment $2\to 3$ whose contribution to the action is Ldt. Considering that $\dot q = 0$ on segment $2\to 3$ and using the relation

$$L = p\dot q - H,$$

we get

$$Ldt\big|_{2\to 3} = -Hdt.$$

By the principle of least action, $\delta S = 0$ and therefore, the first and second paths give the same contribution to the action up to small values $\propto dt$, i.e.

$$S(q,t+dt) = S(q,t) - Hdt,$$

which yields

$$S(q,t+dt) - S(q,t) = -Hdt$$

or

$$\frac{\partial S}{\partial t} = -H. \tag{46.2}$$

To find $\partial S/\partial q$, we consider the variation of $S(q,t)$ when moving from (q,t) (i.e. *point 2* in Fig. 70) to $(q+dq,t)$ (i.e. *point 4* in Fig. 70). As in the previous case, we equate the integral

$$S(q+dq,t) = \int_1^4 Ldt$$

along the curve c and along the 'trial' path, consisting of a segment $1\to 2$ of the curve a and a small segment $2\to 4$. Strictly speaking, the path $1\to 2\to 4$ should not be 'eligible' in the principle of least action, since on the segment $2\to 4$ the quantity $\dot q$ turns to infinity.

But the dependence $q(t)$, which is arbitrarily close and sufficiently smooth, is admissible. Therefore, the relations

$$dt \to 0, \quad \dot{q} \to \infty, \quad \dot{q}dt = dq$$

can be accepted for the segment $2 \to 4$. This segment's contribution to the action of

$$Ldt\big|_{2\to 4} = (p\dot{q} - H)dt = pdq - Hdt$$

is reducible to pdq. Thus,

$$S(q + dq, t) = S(q, t) + pdq,$$

whence $\partial S/\partial q = p$. Using the same approach in case of several degrees of freedom, we obtain

$$\frac{\partial S}{\partial q_i} = p_i. \tag{46.3}$$

Similarly, one may consider the action as a function of the coordinates $q_i^{(1)}$ and the time t_1 at the start of the motion. Thus, the final expression is

$$dS(q, t, q^{(1)}, t_1) = \sum_i p_i dq_i - Hdt - \sum_i p_i^{(1)} dq_i^{(1)} + H^{(1)} dt_1. \tag{46.4}$$

46.2 *Motion of a system as a canonical transformation*

The function $S(q, t, q^{(1)}, t_1)$ can only be determined once the law of motion $q_i(t)$ is known. However, the existence of this function permits certain general conclusions regarding the motion of the system as specified by Hamiltonian equations. For example, it can be concluded that *the transformation from the initial values of coordinates and momenta to their values after time τ is a canonical transformation.*[7] The corresponding generating function is given by:

$$F(q^{(1)}, q, t) = -S(q, t + \tau, q^{(1)}, t), \tag{46.5}$$

where $q^{(1)}$ is considered as the old coordinates and q as the new coordinates. If we account for (4), it becomes clear that under this transformation there is also a transition from the old Hamiltonian $H^{(1)}$ to the new Hamiltonian H.

46.3 *Proof of the Noether's theorem*

The aforementioned properties of $S(q, t, q^{(1)}, t_1)$ enable the easy proof of the Noether's theorem, the formulation of which was provided in § 17.3. Let generalized coordinates

[7] This fact was already noted earlier (see Eq. (43.19)).

$q_i = g_i(t), i = 1, \ldots, s$ describe the actual motion of the system. The action calculated along this trajectory is a function of the initial and final coordinates and time given by:

$$S_{12} \equiv S(q^{(2)}, t_2, q^{(1)}, t_1) = \int_1^2 L\left(g(t), \frac{dg(t)}{dt}, t\right) dt.$$

Since the form of the action remains the same when passing to the variables q_i', t', the equations $q_i' = g_i(t')$ also describe the actual motion of the system. Expressed in terms of q_i, t up to the first order in ε, these equations take the form:

$$q_i + \delta q_i = g_i(t + \delta t),$$

where

$$\delta q_i = \varepsilon f_i(q(t), t), \quad \delta t = \varepsilon h(q(t), t).$$

Consequently, the corresponding action is given by:

$$S_{1'2'} = \int_{1'}^{2'} L\left(g(t'), \frac{dg(t')}{dt'}, t'\right) dt' = S(q^{(2)} + \delta q^{(2)}, t_2 + \delta t_2, q^{(1)} + \delta q^{(1)}, t_1 + \delta t_1).$$

Small changes in the coordinates and time at the start and end of the motion when transitioning from the trajectory $g(t)$ on the segment 12 to the trajectory $g(t')$ on the segment $1'2'$ result in a change in action (4):

$$S_{1'2'} - S_{12} = \sum_i p_i(t_2)\delta q_i^{(2)} - E(t_2)\delta t_2 - \sum_i p_i(t_1)\delta q_i^{(1)} + E(t_1)\delta t_1. \tag{46.6}$$

Note that energy $E(t)$, coordinates $q_i(t)$, and momenta $p_i(t)$ are considered as functions of time on the true trajectory of the system.

On the other hand, according to the theorem's condition, $S_{12} = S_{1'2'}$, hence we obtain the integral of motion:

$$E(t) \cdot h(q(t), t) - \sum_{i=1}^{s} p_i(t) \cdot f_i(q(t), t) = \text{const.} \tag{46.7}$$

§ 47 The Liouville theorem

47.1 *Invariance of a phase volume with respect to canonical transformations*

Imagine a closed $(2s-1)$-dimensional surface that divides a $2s$-dimensional phase space with coordinates $p_1, \ldots, p_s, q_1, \ldots, q_s$ into some region Ω. The integral

$$\Gamma = \int_\Omega dp_1 \ldots dp_s dq_1 \ldots dq_s$$

is called *the phase volume* of this region. Under an arbitrary transformation from the variables q and p to the variables Q and P, the region Ω is transformed into Ω' in terms of the new variables, and the above integral takes the form

$$\Gamma = \int_{\Omega'} D\, dP_1 \dots dP_s dQ_1 \dots dQ_s$$

where

$$D = \frac{\partial(p,q)}{\partial(P,Q)}$$

is the Jacobian of this transformation. Here and below, the letter p denotes the set of variables $p_1, \dots, p_s$, and so do q, P, and Q. On the other hand, the phase volume of the Ω' region is equal to

$$\Gamma' = \int_{\Omega'} dP_1 \dots dP_s dQ_1 \dots dQ_s.$$

We will now prove that if the transformation is canonical, then $D=1$, and the phase volume does not depend on the choice of coordinates, that is $\Gamma' = \Gamma$.

We perform the coordinate transformation in two stages. Firstly, we shift from the variables p and q to the variables P and q, and then to the variables P and Q. The Jacobian of the complete transformation is the product of the Jacobians of the successive stages:

$$D = \frac{\partial(p,q)}{\partial(P,q)} \frac{\partial(P,q)}{\partial(P,Q)}.$$

For each of the factors, we can omit variables that do not change during the transformation by passing to matrices of half the size. Additionally, taking into account that the Jacobians for mutually inverse transformations are inverse and passing to the inverse Jacobian in the second factor, we obtain

$$D = \left.\frac{\partial(p)}{\partial(Q)}\right|_{q=\text{const}} \Bigg/ \left.\frac{\partial(q)}{\partial(Q)}\right|_{P=\text{const}}. \tag{47.1}$$

Using the fact that the final transformation $q, p \to Q, P$ is canonical, we assume that it is given by the generating function $\Phi(q, P)$. By expressing the matrix elements of the Jacobians in the numerator and denominator through Φ, we obtain

$$\frac{\partial p_i}{\partial P_k} = \frac{\partial \Phi}{\partial q_i \partial P_k}, \quad \frac{\partial Q_i}{\partial q_k} = \frac{\partial \Phi}{\partial q_k \partial P_i}.$$

The matrices in the numerator and denominator of equation (1) are mutually transposed, and their determinants are equal; therefore, $D=1$, which is what we needed to prove.

Let each point of the selected phase volume move according to the equations of motion of a given mechanical system. As demonstrated in § 46.2, this motion can be considered as a canonical transformation. This leads to the *Liouville theorem*, which is important in applications: when a Hamiltonian system moves, the phase volume does not change. In other words, the motion of phase space points is similar to the motion of particles in an incompressible fluid.[8] Examples that illustrate the motion of points representing a state of a system in phase space are discussed in [3], problem 11.25f.

47.2 *Focusing lens*

Consider the motion of a group of particles passing through an electrostatic lens. In this case, important qualitative conclusions regarding the particle motion can be drawn thanks to the Liouville theorem, even without knowing the details of the motion. Suppose a group of charged particles with the same momentum projection onto the z-axis and a small spread for both transverse coordinate x and transverse momentum p_x moves on the left in Fig. 71 along the z-axis. For simplicity, we consider only the particles whose trajectories lie in the xz-plane.

In a narrow layer in the vicinity of the point $z=0$, the particles are affected by electric fields that only deflect them in the xz-plane. This layer is called the electrostatic lens, and we assume that the fields are weak enough to make the deflection angles $|\theta_x| = |p_x/p_z| \ll 1$, where p_z is approximately constant for each particle. Is it possible to create a field configuration that reduces the spread of particles in both the transverse coordinate and transverse momentum, i.e. for the angles θ_x?

The Liouville theorem forbids such focusing. Assuming $p_z = \text{const}$ and substituting $z = p_z t/m$, the Hamiltonian for the one-dimensional transverse motion can be obtained from the Hamiltonian of the system $H(x, z, p_x, p_z, t)$. The initial spread of particles over x, p_x corresponds to some area in the phase plane. A simultaneous focusing of a coordinate and momentum after the passage of the lens would mean a decrease in this area, which is impossible due to the Liouville theorem.

Now, consider the transformation of the area in the phase plane during the passage through a field configuration acting as a focusing lens, i.e. which compresses a wide beam along the x-axis into a narrow beam. Figure 72 shows the phase plane x, p_x for the transverse motion. The ellipse *1* represents the area where the images of the particles of the incident beam are concentrated in front of the lens. After passing through the lens, the spread of the angles sharply increases with the same beam size (see ellipse *2*). Then, there is a free motion in the transverse direction leading to the beam compression (ellipse *3*). However, the beam expands again after passing through the lens (ellipse *4*), and the spread of the angles remains unchanged. The dependence of the particle concentration in the beam on z can be described using the formulae from [3], problem 11.25f.

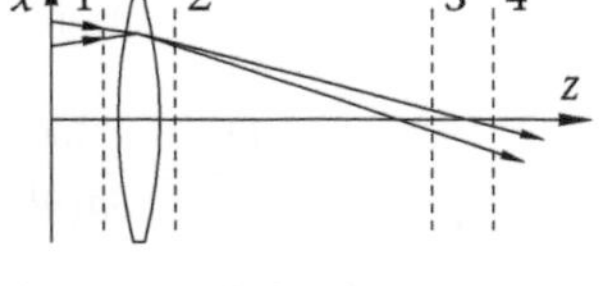

Figure 71 Particle motion in the area of the focusing lens: the dotted lines correspond to the times when the group of particles is in front of lens (*line 1*), immediately after it (*line 2*), in the focus area (*line 3*), and after it (*line 4*)

[8] For Cartesian coordinates, the Liouville theorem is valid not only in phase space but also in the space of coordinates and velocities.

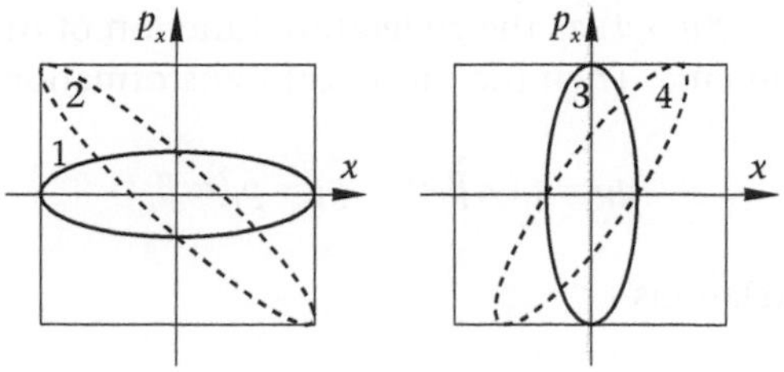

Figure 72 Areas of the phase space occupied by a particle beam in different times: *1–2–3–4*

Problem

47.1. What are the changes over time in the volume, the momentum-space volume, and the phase-space volume occupied by a group of particles moving freely along the x-axis? At time $t = 0$, the particle coordinates lie within the interval of $x_0 < x < x_0 + \Delta x_0$, and their momenta are within the range of $p_0 < p < p_0 + \Delta p_0$.

a) Do the same for particles which move along the x-axis between two walls. Collisions with the walls are absolutely elastic. The particles do not interact with one another.

b) Do the same for a group of harmonic oscillators.

c) Do the same for a group of harmonic oscillators with friction.

§ 48 The Hamilton–Jacobi equation

48.1 *The Hamilton–Jacobi equation. Separation of the variables*

The action $S(q, t, q^{(1)}, t_1)$ satisfies the equation which is easy to obtain if in relation (46.2)

$$\frac{\partial S}{\partial t} = -H(q_1, q_2, \dots, q_s, p_1, p_2, \dots, p_s, t)$$

we replace p_i with its expression $p_i = \partial S/\partial q_i$ from Eq. (46.3). As a result, we get the *Hamilton–Jacobi equation*

$$\frac{\partial S}{\partial t} + H\left(q_1, \dots, q_s, \frac{\partial S}{\partial q_1}, \dots, \frac{\partial S}{\partial q_s}, t\right) = 0 \tag{48.1}$$

similar to the eikonal equation for light rays. It can be used to solve the problem of the motion of mechanical systems. To do this, we need to find the complete integral of Eq. (1), which is a function that satisfies this equation, with s arbitrary parameters $\alpha_1, \dots, \alpha_s$ (the insignificant additive $(s+1)$-th constant α_{s+1} may be ignored):

$$S = f(q_1, \dots, q_s, \alpha_1, \dots, \alpha_s, t) + \alpha_{s+1}. \tag{48.2}$$

Then finding $q_i(t)$ and $p_i(t)$ is reduced to solving only algebraic equations.

To show this, we take $f(q, \alpha, t)$ as the generating function of the canonical transformation and α_i as new momenta. Then the canonical transformation

$$q_i = q_i(\alpha, \beta, t), \quad p_i = p_i(\alpha, \beta, t) \tag{48.3}$$

is determined by the relations

$$\frac{\partial f}{\partial q_i} = p_i, \quad \frac{\partial f}{\partial \alpha_i} = \beta_i, \tag{48.4}$$

where β_i denotes the new coordinates. The new Hamiltonian

$$H'(\alpha, \beta, t) = H + \frac{\partial f}{\partial t}$$

according to (1), (2) turns out to be identically equal to zero: $H'(\alpha, \beta, t) = 0$. Due to equations

$$\dot{\alpha}_i = -\frac{\partial H'}{\partial \beta_i} = 0, \quad \dot{\beta}_i = -\frac{\partial H'}{\partial \alpha_i} = 0$$

new canonical variables turn out to be integrals of motion, while Eqs. (3) determine the law of the motion. Note that the parameters α and β are sufficient to satisfy the initial conditions. Therefore, knowing the complete integral of the Hamilton–Jacobi equation means that the further computation of $q_i(t)$ and $p_i(t)$ is purely an algebraic problem.

One may wonder how to find the complete integral and even whether it is always possible. The complete integral can be found by carrying out separation of variables for cases when the problem of integrating the equations of motion can be reduced to quadratures, which are regarded as exceptions. Cases that are of greatest physical interest are listed in [1], § 48. Such are, among others, the motion of particles (i) in the field of a pair of fixed Coulomb centres, and (ii) in the Coulomb field combined with a uniform field. At the same time, it is often a good choice of curvilinear coordinates which ensures a successful solution of the problem.

So, the complete integral of the Hamilton–Jacobi equation is obtained via separation of variables. Yet, if the Hamiltonian does not explicitly depend on time, S can be found in the form:

$$S(q_1, \ldots, q_s, t) = -Et + S_0(q_1, \ldots, q_s), \tag{48.5}$$

where $S_0(q_1, \ldots, q_s)$ is called the *abbreviated action*. For that, we get the equation

$$H\left(q_1, \ldots, q_s, \frac{\partial S_0}{\partial q_1}, \ldots, \frac{\partial S_0}{\partial q_s}\right) = E. \tag{48.6}$$

Next, we present a function of many variables $S_0(q_1, \ldots, q_s)$ as a sum of functions of one variable

$$S_0(q_1, \ldots, q_s) = \sum_{i=1}^{s} S_i(q_i). \tag{48.7a}$$

Substituting this expression into Eq. (6), instead of the equation in partial derivatives we obtain the following equation

$$H\left(q_1, \dots, q_s, \frac{dS_1(q_1)}{dq_1}, \dots, \frac{dS_s(q_s)}{dq_s}\right) = E.$$

It contains only the total derivatives of the functions $dS_i(q_i)/dq_i$. The solution to this equation has the form

$$S_i(q_i) = \int p_i(q_i)\, dq_i, \tag{48.7b}$$

where $p_i(q_i)$ is a known function of the corresponding coordinate. It is clear that Eq. (6) has a solution of this form by no means always, as already discussed above.

48.2 *Motion of a relativistic particle in the field* $U(r) = -\frac{\alpha}{r}$

Let us consider in more detail the application of the method of the variable separation using the motion of a relativistic particle in the Coulomb field as an example.

In this case, the Hamiltonian has the form (38.12):

$$H(\mathbf{r}, \mathbf{p}, t) = \sqrt{\mathbf{p}^2c^2 + m^2c^4} - \frac{\alpha}{r},$$

where the relativistic momentum is

$$\mathbf{p} = \frac{m\mathbf{v}}{\sqrt{1 - (v/c)^2}}.$$

When moving in a central field, relativistic energy $\mathcal{E}$ and angular momentum $\mathbf{M}$ are conserved:

$$\mathcal{E} = \sqrt{\mathbf{p}^2c^2 + m^2c^4} - \frac{\alpha}{r}, \quad \mathbf{M} = [\mathbf{r}, \mathbf{p}] = \frac{m\,[\mathbf{r}, \mathbf{v}]}{\sqrt{1 - (v/c)^2}}.$$

This shows that the motion of the particle occurs in a plane perpendicular to the constant vector $\mathbf{M}$. We introduce in this plane the polar coordinates r and φ, and the Lagrangian in these coordinates is

$$L = -mc^2\sqrt{1 - \frac{\dot{r}^2 + r^2\dot{\varphi}^2}{c^2}} + \frac{\alpha}{r}.$$

The corresponding generalized momenta

$$p_r = \frac{\partial L}{\partial \dot{r}} = \frac{m\dot{r}}{\sqrt{1 - (v/c)^2}}, \quad p_\varphi \equiv M = \frac{\partial L}{\partial \dot{\varphi}} = \frac{mr^2\dot{\varphi}}{\sqrt{1 - (v/c)^2}}$$

are connected by the relation (it has the same form as in the non-relativistic case—see (10.9))

$$\mathbf{p}^2 = p_r^2 + \frac{M^2}{r^2}.$$

It follows that

$$p_r(r) = \pm\sqrt{\frac{1}{c^2}\left(\mathcal{E} + \frac{\alpha}{r}\right)^2 - \frac{M^2}{r^2} - m^2c^2}.$$

Seeking a solution to the Hamilton–Jacobi equation for a abbreviated action in the form (7):

$$S_0(r, \varphi) = \int p_\varphi(\varphi)\, d\varphi + \int p_r(r)\, dr$$

we obtain as a result the complete integral of the Hamilton–Jacobi equation

$$S(r, \varphi, \mathcal{E}, M, t) = -\mathcal{E}\, t + M\varphi + \int p_r(r)\, dr,$$

where the quantities $\mathcal{E}$ and M play the role of parameters $\alpha_{1,2}$.

The equation $\partial S/\partial\alpha_1 = \beta_1$ or $\partial S/\partial\mathcal{E} = \text{const}$ defines the dependency $r(t)$:

$$t = \frac{1}{c^2}\int \left(\mathcal{E} + \frac{\alpha}{r}\right)\frac{dr}{p_r(r)}, \tag{48.8}$$

while the equation $\partial S/\partial M = \text{const}$ or

$$\varphi = \int \frac{M}{r^2}\frac{dr}{p_r(r)} \tag{48.9}$$

determines the trajectory of the particle. Now, let us compare equation (9) with equation (5.2) for the trajectory of a non-relativistic particle in the field

$$U(r) = -\frac{\alpha}{r} + \frac{\beta}{r^2}.$$

It can be seen that Eq. (9) coincides with Eq. (5.2) under the replacements

$$m \to \frac{\mathcal{E}}{c^2},\quad E \to \frac{\mathcal{E}^2 - m^2c^4}{2\mathcal{E}},\quad \beta \to -\frac{\alpha^2}{2\mathcal{E}}. \tag{48.10}$$

In other words, Eq. (9) corresponds to the motion of the non-relativistic particle in the Coulomb field $U(r) = -\alpha/r$ with perturbation in the form of an additional central field of attraction

$$\delta U(r) = -\frac{\alpha^2}{2\mathcal{E}\, r^2}. \tag{48.11}$$

When $M > \alpha/c$, the result for the case under consideration can be expressed in a form similar to (5.4):

$$r = \frac{\tilde{p}}{1 + \tilde{e}\cos(\gamma\varphi)}, \tag{48.12}$$

where we introduce the notations:

$$\gamma = \sqrt{1 - \frac{\alpha^2}{c^2 M^2}}, \quad \tilde{p} = \frac{c^2 M^2 - \alpha^2}{\mathcal{E}\alpha}, \quad \tilde{e} = \sqrt{1 + \frac{(\mathcal{E}^2 - m^2 c^4)(c^2 M^2 - \alpha^2)}{\mathcal{E}^2 \alpha^2}}. \tag{48.13}$$

When $\mathcal{E} < mc^2$ (and $\tilde{e} < 1$), the trajectory represents the finite motion of the particle and generally is not closed (see Fig. 8*b*). With one radial oscillation of the particle, the polar angle changes by:

$$\Delta\varphi = \frac{2\pi}{\gamma}. \tag{48.14}$$

If we apply the above formulae to the motion of planets in the solar system, the associated change corresponds to a perihelion shift of the planets with an angle of

$$\delta\varphi = \frac{2\pi}{\gamma} - 2\pi. \tag{48.15a}$$

The non-relativistic limit of this expression applies to the motion of planets where $v \ll c$ or $mc^2 - \mathcal{E} \approx \alpha/(2a) \ll mc^2$:

$$\delta\varphi \approx \pi \frac{\alpha/a}{mc^2(1 - e^2)}, \tag{48.15b}$$

where a and e are the semi-major axis and the eccentricity of the planet. For the planet Mercury, the perihelion shift is numerically equal to

$$\delta\varphi_{\mathrm{SRT}} \approx 7.16355'' \ \text{ per century.} \tag{48.16}$$

The observed precession of the Mercury's perihelion (after eliminating the influence other planets) is

$$\delta\varphi_{\mathrm{observ}} = (42.9777 \pm 0.0050)'' \ \text{ per century.} \tag{48.17}$$

This value of $\delta\varphi_{\mathrm{observ}}$ exhibits an obvious contradiction with the prediction (16) of special relativity theory but shows excellent agreement with the prediction of general relativity theory (see [10], § 101):

$$\delta\varphi_{\mathrm{GRT}} = 6\delta\varphi = 42.9813'' \ \text{ per century.} \tag{48.18}$$

48.3 *Opto-mechanical analogy*

It is known that the propagation of a light wave travelling through a transparent medium with refractive index $n(\mathbf{r})$ can be described as the motion of wave surfaces represented by surfaces of identical phases $\Psi(\mathbf{r}, t) = \text{const}$. In the approximation of geometric optics, the phase $\Psi(\mathbf{r}, t)$ satisfies the *eikonal equation*:

$$(\nabla\Psi)^2 = \frac{n^2}{c^2}\left(\frac{\partial\Psi}{\partial t}\right)^2,$$

where c indicates the velocity of light in a vacuum (see [9], § 85).

A reminder: as per (38.12a), the Hamiltonian for a photon appears as:

$$H(\mathbf{r}, \mathbf{p}) = \frac{c}{n(\mathbf{r})} |\mathbf{p}|.$$

The corresponding Hamilton–Jacobi equation

$$\frac{\partial S}{\partial t} = \frac{c}{n(\mathbf{r})} |\nabla S|$$

can be easily converted to

$$(\nabla S)^2 = \frac{n^2(\mathbf{r})}{c^2} \left(\frac{\partial S}{\partial t}\right)^2.$$

This equation coincides with the eikonal equation if we assume that the function $S(\mathbf{r}, t)$ is proportional to the phase $\Psi(\mathbf{r}, t)$. In other words, in photon mechanics, the role of wave surfaces is played by the surfaces $S(\mathbf{r}, t) = \text{const}$, where $S(\mathbf{r}, t)$ is the action as a function of coordinates and time.

Problem

48.1. Determine trajectories and laws of motion for particles scattered in the field

$$U(r, \theta, \varphi) = \frac{a \cos\theta}{r^2}$$

and falling into the centre of this field. Express the trajectory in terms of quadratures and analytically for $E\rho^2 \ll a$. Assume the particle velocity before the scattering to be parallel to the z-axis.

§ 49 Angle and action variables

If a mechanical system allows for separation of the variables in the Hamilton–Jacobi equation and the system makes a finite motion, then special canonical variables $w_1, \ldots, w_s, I_1, \ldots, I_s$ known as *angle and action variables* can be introduced. These variables can give a unified and simple description of all such systems. The variables are useful in studying mechanical systems derived from a more or less significant modification of the original, and that do not allow the separation of variables. Modern research in non-linear mechanics often formulates the problem in terms of these variables. The discussion begins with systems for which the Hamiltonian does not explicitly depend on time: $\partial H/\partial t = 0$.

49.1 *Systems with one degree of freedom*

Consider a system with one degree of freedom, performing periodic motion, with Hamiltonian $H(q, p)$. One can introduce new canonical variables (coordinate w and momentum I) for such systems. In these coordinates, the motion of the system is extremely simple, even if the motion in the initial variables q and p was quite complicated. The momentum I is conserved, while the coordinate w grows linearly with time, corresponding to the free motion of a point in these variables. In this case, the new description of the system will be exactly the same for a wide variety of systems as the entire individuality of the original system will be 'hidden' in formulae that relate the old and new variables. We consider an instructive example before proceeding to the general case.

Example

For a harmonic oscillator with the Hamiltonian

$$H(x,p) = \frac{p^2}{2m} + \frac{1}{2}m\omega^2x^2, \tag{49.1}$$

new variables can be introduce as

$$x = \sqrt{\frac{2I}{m\omega}}\sin w, \quad p = \sqrt{2m\omega I}\cos w \tag{49.2}$$

using the generating function (11) given below. It can be easily verified that the transformation is canonical as

$$\{p,\, x\}_{I,w} = 1.$$

The new Hamiltonian

$$H'(w,I) = H(x(w,I),\, p(w,I)) = \omega I$$

does not depend on the w coordinate and, therefore, I is conserved and equals

$$I = \frac{E}{\omega} = \text{const}, \tag{49.3}$$

since $H = E = \text{const}$.

The Hamiltonian equation for the new coordinate

$$\dot{w} = \frac{\partial H'}{\partial I} = \omega \tag{49.4}$$

leads to the trivial law of motion:

$$w(t) = \omega t + w_0. \tag{49.5}$$

Figure 73 shows the phase trajectory in the initial variables x and p, which is in the shape of ellipses, and rectilinear phase trajectory in angle and action variables w and I.

Now, consider the one-dimensional motion in the general case. The separation of variables in this case is a trivial matter. Given the equality

$$H(q,p) = E, \tag{49.6}$$

we express p as a function of q and E, which yields the abbreviated action

$$S_0(q,E) = \int_{q_0}^{q} p(q,E)\, dq. \tag{49.7}$$

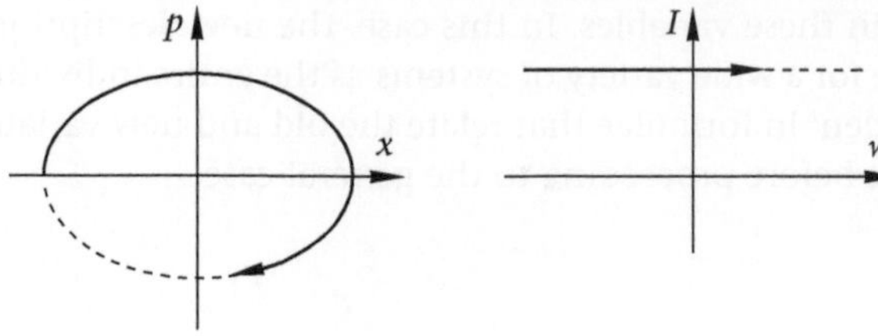

Figure 73 Phase trajectories in initial variables x and p as ellipses, and in angle and action variables w and I as straight lines

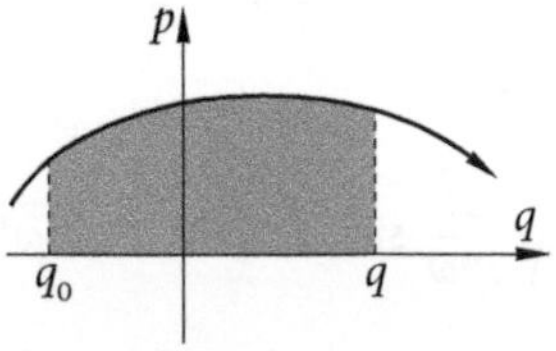

Figure 74 The geometric meaning of the abbreviated action is the highlighted area on the phase plane

Next, we define I as

$$I(E) = \frac{1}{2\pi} \oint p(q, E)\, dq, \tag{49.8}$$

where the integral is taken over the period of motion. For example, for the pendulum, one can imagine that, depending on the energy, the period of motion is either a period of oscillation or a full turnover period. Geometrically, $S_0(q, E)$ represents the highlighted area on the phase plane as shown in Fig. 74, and $2\pi I$ corresponds to the area bounded by the phase trajectory for one period of motion.

It is important to note that $S_0(q, E)$ is a many-valued function of the coordinate, increasing over the period by $2\pi I$. Therefore, after n periods of motion, the increment is

$$\Delta S_0 = 2\pi n I. \tag{49.9}$$

By expressing $E(I)$ from (8) and substituting into (7), we obtain the function

$$\Phi(q, I) = S_0(q, E(I)), \tag{49.10}$$

which we take as the generating function of the canonical transformation, where I is a new momentum.[9] The transformation from the old to the new canonical variables is

[9] In the above example of the harmonic oscillator, the momentum is $p(x, E) = \sqrt{2mE - (m\omega x)^2}$, while abbreviated action is equal to

$$S_0(x, E) = \int_{x_0}^{x} p(x, E)\, dx = \frac{1}{2} x\sqrt{2mE - (m\omega x)^2} + \frac{E}{\omega} \arcsin\left(\sqrt{\frac{m}{2E}}\, \omega x\right) + 2\pi n \frac{E}{\omega} + \text{const.}$$

given by relations of the form:

$$p = \frac{\partial \Phi(q, I)}{\partial q}, \quad w = \frac{\partial \Phi(q, I)}{\partial I}. \tag{49.12}$$

The new Hamiltonian is already known; it does not depend on w coordinates:

$$H'(I, w) = E(I). \tag{49.13}$$

The Hamiltonian equations

$$\dot{I} = -\frac{\partial H'}{\partial w} = -\frac{\partial E}{\partial w} = 0, \quad \dot{w} = \frac{\partial E}{\partial I} \tag{49.14}$$

lead to the law of motion for new variables

$$I = \text{const}, \quad w = \frac{\partial E}{\partial I} t + w_0. \tag{49.15}$$

The quantity $\partial E/\partial I$ can be determined using Eq. (8):

$$\frac{\partial I}{\partial E} = \frac{1}{2\pi} \oint \frac{\partial p(q, E)}{\partial E}\, dq. \tag{49.16}$$

To obtain the derivative $\partial p(q, E)/\partial E$, we substitute $p = p(q, E)$ into Eq. (6) and differentiate the resulting identity, $H(q, p(q, E)) = E$, with respect to E. This yields

$$\frac{\partial H}{\partial p}\frac{\partial p}{\partial E} = 1,$$

where

$$\frac{\partial p}{\partial E} = \frac{1}{\partial H/\partial p} = \frac{1}{\dot{q}}, \tag{49.17}$$

and thus

$$\frac{\partial I}{\partial E} = \frac{1}{2\pi} \oint \frac{dq}{\dot{q}} = \frac{T}{2\pi} = \frac{1}{\omega}, \tag{49.18}$$

where $T = 2\pi/\omega$ is the period of motion and ω is the frequency. In a general case, the frequency ω depends on the energy, i.e. on I. As a result, we have

$$I = \text{const}, \quad w = \omega(I)\, t + w_0. \tag{49.19}$$

The angle variable w, as well as the abbreviated action S_0, is a many-valued function of the coordinate q. For the period of motion, the variable w is increased by 2π (cf. (9)). On the other hand, variables q and p are the periodic functions of w.

Hence the generating function reads

$$\Phi(x, I) = \frac{1}{2} x\sqrt{2m\omega I - (m\omega x)^2} + I \arcsin\left(\sqrt{\frac{m}{2\omega I}}\, \omega x\right) + 2\pi n I + \text{const}. \tag{49.11}$$

49.2 *Systems with many degrees of freedom*

For a system with many degrees of freedom and a Hamiltonian that does not explicitly depend on time ($\partial H/\partial t = 0$), we can introduce angle and action variables, provided that the variables are separated. This is achieved by using the same procedure as in the one-dimensional case. However, the function S_0 in Eq. (7) should be replaced with the functions S_i from Eq. (48.7). Note that each of the variables I_i turns out to be a function of s arbitrary constants α_j, including the energy E. The new Hamiltonian is given by

$$H' = E(I_1, \dots, I_s), \tag{49.19}$$

while pairs of canonically conjugate coordinates and momenta, q_i and p_i, respectively, are functions of their 'own' w_i and all variables I_j. Moreover, parameters α_i can also be expressed in terms of I_j:

$$\alpha_i = \alpha_i(I_1, \dots, I_s). \tag{49.20}$$

The variables I_i play a special role in the transition to quantum mechanics. In the so-called quasi-classical approximation, they can take only a discrete series of values (*the Bohr and Sommerfeld's quantization rules*):

$$I_i = n_i\, \hbar,$$

where n_i is an integer and $\hbar$ is Planck's constant or quantum of action.

49.3 *Hamiltonian function explicitly depending on time*

Let us consider a mechanical system with one degree of freedom, where the Hamiltonian function depends on a time-varying parameter $\lambda = \lambda(t)$. In this case, the Hamiltonian function $H = H(q, p, \lambda)$ explicitly depends on time through λ. This results in the non-conservation of the system's energy E. The canonical variables w and I can still be obtained using the same formulae (8)–(10) and (12), but with the inclusion of λ. We can determine the momentum $p = p(q, E, \lambda)$ by equating the Hamiltonian with E for a fixed λ:

$$H(q, p, \lambda) = E. \tag{49.21}$$

After that, we can calculate the abbreviated action $S_0(q, E, \lambda)$ and the generating function $\Phi(q, I, \lambda)$.

The new Hamiltonian function according to (43.16) is

$$H'(w, I, \lambda) = E(I, \lambda) + \frac{\partial \Phi(q, I, \lambda)}{\partial t} = E(I, \lambda) + \dot{\lambda}\, \frac{\partial \Phi(q, I, \lambda)}{\partial \lambda}. \tag{49.22}$$

In this expression, the variable q in the function $\partial \Phi(q, I, \lambda)/\partial \lambda$ should be replaced to $q = q(w, I)$ found from Eqs. (12). Let us denote the resulting function as

$$\Lambda(w, I, \lambda) = \left.\frac{\partial \Phi(q, I, \lambda)}{\partial \lambda}\right|_{q=q(w,I)}. \tag{49.23}$$

It becomes immediately evident that the function $\Lambda(w, I, \lambda)$ is a single-valued function[10] of the variable w (unlike $\Phi(q, I, \lambda)$). Indeed, the differentiation of the function $\Phi(q, I, \lambda)$ with respect to λ is done at a fixed value of I, so the addition $2\pi I$, that occurs after each oscillation period (see formulae (9), (10)), disappears.

We finally obtain the new Hamiltonian

$$H'(I, w, \lambda) = E(I, \lambda) + \dot{\lambda}\Lambda(I, w, \lambda) \tag{49.24}$$

and the Hamiltonian equations for new variables (cf. (14), (18))

$$\dot{I} = -\dot{\lambda}\frac{\partial \Lambda}{\partial w}, \quad \dot{w} = \omega + \dot{\lambda}\frac{\partial \Lambda}{\partial I}. \tag{49.25}$$

These equations are convenient for constructing the perturbation theory when $\dot{\lambda}$ turns out to be small (see § 50.3).

Example

Let us consider a harmonic oscillator with a time-dependent frequency $\omega(t)$ represented by the Hamiltonian function

$$H(p, x, \omega) = \frac{p^2}{2m} + \frac{1}{2}m\omega^2(t)x^2. \tag{49.26}$$

Here, the parameter λ is $\omega(t)$. Using Eqs. (11) and (2) in the same manner as before, we can determine the generating function $\Phi(q, I, \lambda)$ and the relationship between the old and new variables. According to (22)–(24), the new Hamiltonian is equal to

$$H'(w, I, \omega) = \omega I + \dot{\omega}\frac{\partial \Phi(q, I, \omega)}{\partial \omega} =$$

$$= \omega I + \frac{\dot{\omega}}{2\omega}\sqrt{2m\omega I - (m\omega x)^2} = \omega I + \frac{\dot{\omega}}{2\omega} I \sin 2w, \tag{49.27}$$

and the equations of motion are

$$\dot{I} = -\frac{\dot{\omega}(t)}{\omega(t)} I \cos 2w, \quad \dot{w} = \omega(t) + \frac{\dot{\omega}(t)}{2\omega(t)}\sin 2w. \tag{49.28}$$

If the frequency ω changes slowly and smoothly, the right side of the equation for the action is small due to the oscillating factor $\cos 2\omega$ at the small parameter $\dot{\omega}(t)$. When averaged over the period of motion, the right side disappears, resulting in the action to be on average conserved (for more details, see [3], problem 13.11).

[10] If the parameter λ does not vary with time, then the motion of the system is periodic and the single-valued function $\Lambda(w, I, \lambda)$ is a periodic function of the variable w.

§ 50 Adiabatic invariants

50.1 *Setup of the problem and result*

Consider a mechanical system with one degree of freedom that performs oscillations while its parameter λ varies *adiabatically slowly*. Here, 'adiabatically slowly' means that the variation of λ is slow and smooth. In other words, the change of this parameter $\sim \dot{\lambda} T$ over the period of oscillations T is small compared to the parameter λ itself, and also the rate of this change $\ddot{\lambda}$ is small compared to $\dot{\lambda}/T$:

$$\dot{\lambda} T \ll \lambda, \quad \ddot{\lambda} T \ll \dot{\lambda}. \tag{50.1}$$

Let $H(q, p, \lambda)$ be the Hamiltonian, and if λ is constant, then energy E is conserved and the trajectory on the phase plane is closed. The area enclosed by this trajectory is determined by the equation

$$H(q, p, \lambda) = E. \tag{50.2}$$

If $p = p(q, E, \lambda)$ is the solution to this equation, then the area enclosed on the phase plane by this trajectory is equal to

$$S(E, \lambda) = \oint p(q, E, \lambda)\, dq. \tag{50.3}$$

If λ varies slowly, the energy E is no longer conserved, but its average value over the period

$$\langle E \rangle \equiv \frac{1}{T} \int_0^T E(t) dt \tag{50.4}$$

does change slowly and is determined significantly by the variation of the parameter $\lambda(t)$. We can then form a function from $E(t)$ and $\lambda(t)$ such that the average value of the combination will be constant. Such a combination is called an *adiabatic invariant* and, as will be shown further, corresponds to the canonical variable—the action I. As mentioned earlier, the action, up to a factor, is equivalent to the area enclosed by the phase trajectory

$$S(E, \lambda) = 2\pi I(E, \lambda)$$

and calculated for given (fixed in current time) values of E and λ:

$$I(E, \lambda) = \frac{S(E, \lambda)}{2\pi} = \frac{1}{2\pi} \oint p(q, E, \lambda)\, dq. \tag{50.5}$$

In § 50.3, we will show that

$$\langle I(E, \lambda) \rangle = \text{const} \tag{50.6}$$

up to values of the order of $\dot{\lambda} T$ inclusive. Here we present the qualitative arguments based on the Liouville theorem. Assume that the parameter $\lambda(t)$ in the initial moment has the value λ and corresponds to the initial energy E. At fixed value of this parameter (and energy) the phase trajectory is closed, confining the phase region of the area S.

However, the energies of individual points within this region differ from E, and the energies of points on the boundary of the area cease to be the phase trajectory when the parameter $\lambda(t)$ varies arbitrarily. When $\lambda(t)$ varies adiabatically slowly, all points of the initial phase trajectories can remain points of the phase trajectory for the new energy E' corresponding to the new value of the parameter λ'. Thus, the adiabatic invariant is conserved.

The two typical examples below will illustrate these concepts.

Example 1

As the first example, we consider the harmonic oscillator with a slowly varying frequency (49.26). In this case, the condition (1) implies that

$$\dot{\omega} \ll \omega^2, \quad \ddot{\omega} \ll \dot{\omega}\omega. \tag{50.7}$$

The phase trajectory of the harmonic oscillator is an ellipse given by equation $H(x, p, \omega) = E$ with semi-axes $\sqrt{2mE}$ and $\sqrt{2E/(m\omega^2)}$ and area $S = 2\pi E/\omega$. The adiabatic invariant is (cf. (49.3))

$$I = \frac{S}{2\pi} = \frac{E}{\omega}, \tag{50.8}$$

which indicates that the energy of the harmonic oscillator varies directly proportionally to its frequency: $E = I\omega$.

Example 2

As the second example, we consider a particle in rectangular potential box whose width $l(t)$ slowly changes (see Fig. 75, where the elastic wall at the point $x=0$ is motionless, and the elastic wall at the point $x = l(t)$ moves slowly away). The oscillation period is $T = 2l/v$, where $v = |\dot{x}|$ is a particle velocity modulus. In this case, the condition (1) implies that

$$\dot{l} \ll v, \quad \ddot{l}l \ll \dot{l}v. \tag{50.9}$$

The phase trajectory of this system is a rectangle with sides l and $2mv = 2\sqrt{2mE}$ (Fig. 76), so the adiabatic invariant is

$$I = \frac{m}{\pi} v l = \frac{\sqrt{2m}}{\pi} \sqrt{E}\, l. \tag{50.10}$$

Thus, the particle energy in such a box is $E \propto 1/l^2$.

Figure 75 Particle between a stationary wall and a slowly receding wall

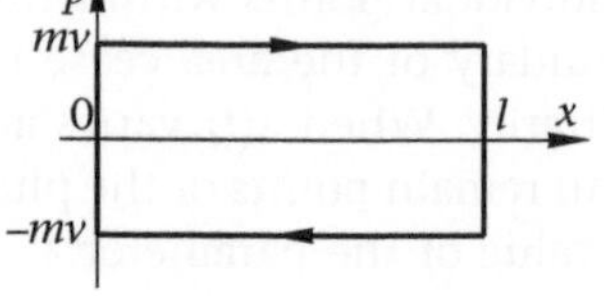

Figure 76 Phase trajectory of the system shown in Fig. 75

This example is relevant to adiabatic processes in gases (see problem 13.14 from [3]) and to the adiabatic approximation in the theory of molecules (see problem 13.10 from [3]).

50.2 *Adiabatic invariant for a particle in a box*

Before proceeding to the proof of result (6), let us consider the simple Example 2 in more detail, which is convenient for demonstrating how the considered values depend on time.

When a particle collides with a fixed wall, its velocity is not changed in magnitude, whereas when a particle collides with a receding wall, its velocity decreases in magnitude by $2\dot{l}$. Let us choose such time Δt, that

$$\frac{2l}{v} \ll \Delta t \ll \frac{l}{\dot{l}}.$$

Such a Δt exists due to condition (9). During this time, there are $v\Delta t/(2l)$ pairs of collisions with walls, and the particle's velocity changes by

$$\Delta v = -v\dot{l}\frac{\Delta t}{l}.$$

The quantities Δv and Δt are small, so their ratio can be considered as an acceleration $\Delta v/\Delta t = dv/dt$. Integrating the resulting equation, we find the result (10): $vl = \text{const}$.

We can also observe how vl changes in detail by examining graphs of $l(t)$ and $v(t)$ (Fig.77*a*, *b*). The product vl is shown in Fig. 77*c*. Value vl oscillates around the approximately constant value of $\langle vl\rangle$, and the oscillation amplitude has a relative value $\sim \dot{l}/v$. A deviation $\langle vl\rangle$ from constant has the highest order of smallness, which is given by

$$\frac{d}{dt}\langle vl\rangle \sim \dot{l}^2.$$

As l increases, the particle's velocity decreases, and there will be a moment when the particle will no longer be able to overtake the receding wall. After this point in time, the velocity of the particle remains constant, while the product vl continues to grow. A violation of the condition of adiabaticity (9) will occur even before that.

50.3 *Conservation of the adiabatic invariant*

Let us prove relation (6). The canonical variable I satisfies Eq. (49.25):

$$\dot{I} = -\dot{\lambda}\frac{\partial\Lambda(w, I, \lambda)}{\partial w}. \qquad (50.11)$$

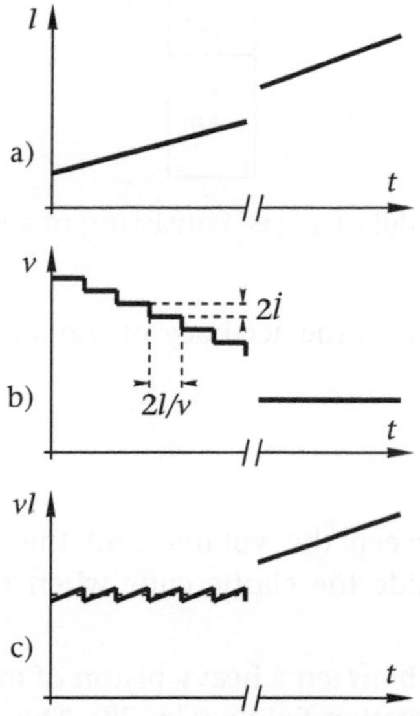

Figure 77 Change of l, v, and vl with time for the system shown in Fig. 75

The right side of this equation contains the small parameter $\dot{\lambda}$. Therefore, this equation can be solved using the method of successive approximations. Considering the time interval of the order of the period of oscillations, in the first approximation in $\dot{\lambda}$, we can substitute variables (49.19)

$$I = \text{const}, \quad w = \omega t + \omega_0, \quad \lambda = \text{const}$$

into the function $\partial\Lambda(w, I, \lambda)/\partial w$, standing at this small factor $\dot{\lambda}$.

In such an approximation

$$\frac{\partial\Lambda(w, I, \lambda)}{\partial w} = \frac{1}{\omega}\frac{d\Lambda}{dt},$$

and therefore Eq. (11) can be rewritten in the form

$$\frac{dI}{dt} = -\frac{\dot{\lambda}}{\omega}\frac{d\Lambda}{dt}. \tag{50.12}$$

Considering that $\ddot{\lambda}T \ll \dot{\lambda}$, we can assume that $\dot{\lambda} = \text{const}$ within the period. Integrating Eq. (11) over the period of motion, we obtain

$$I(t+T) - I(t) = -\frac{\dot{\lambda}}{\omega}\,[\Lambda(t+T) - \Lambda(t)] = 0, \tag{50.13}$$

since in the considered approximation the function Λ is periodic (see footnote 10 to § 49). Within the period, the adiabatic invariant $I(t)$ changes by an amount of the order of $\dot{\lambda}T\Lambda$, but it returns to its original value within a period. It follows from this that the adiabatic invariant oscillates with a small amplitude around a constant value, not deviating from it for many periods. Such dependence $I(t)$ is clearly visible in Fig. 77 (and, as the reader may recall, $I = mvl/\pi$).

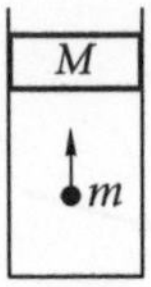

Figure 78 Model of a 'gas' consisting of a single molecule

Note that in a number of cases the accuracy of the adiabatic invariant can be much higher (see [1], § 51).

Problems

50.1. Find a connection between the volume and the pressure of a 'gas' consisting of particles moving inside the elastic cube when the size of the cube is slowly changing.

50.2. A ball of mass m moves between a heavy piston of mass $M \gg m$ and the bottom of the cylinder in the gravitation field g (Fig. 78). The equilibrium distance from the bottom of the cylinder to the piston is equal to X_0. Considering that the velocity of the ball is much more than the piston's velocity, determine the law of motion of the piston, averaged over the period of the ball's motion. Find the frequency of small oscillations of the piston. Assume the collisions to be elastic. (This is a model of a 'gas' consisting of a single molecule.)

50.3. Consider small oscillations of the pendulum in the field of gravity. The length of the pendulum slowly doubles. Find how the maximum deflection angle of the pendulum will change.

50.4. Find the period of the oscillation of an electron along the z-axis in a magnetic trap. The magnetic field of the trap is symmetric with respect to the z-axis, and its components in the cylindrical coordinate system are equal to

$$B_\varphi = 0, \quad B_z = B_z(z), \quad B_r = -\frac{1}{2}\, r B_z'(z),$$

where

$$B_z(z) = B_0\left(1 + \frac{z^2}{a^2}\right).$$

§ 51 Motion of a system with many degrees of freedom. Dynamic chaos

The motion of a system with many degrees of freedom, in contrast to the one-dimensional case, cannot be exhaustively investigated. A natural approach is to consider it in a $2s$-dimensional phase space with coordinates q and p. Of particular interest is the description of the finite motion of the system over an unlimited time.

The motion of multidimensional systems is highly diverse. For systems with separable variables, it is true that all s action variables are integrals of motion:

$$I_i(q_1, \ldots, q_s, p_1, \ldots, p_s) = \text{const}_i, \quad i = 1, 2, \ldots, s, \tag{51.1}$$

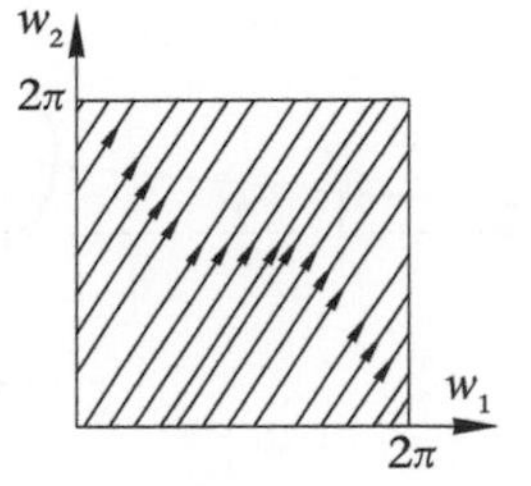

Figure 79 Phase trajectory on the w_1-w_2 plane

therefore, in the phase space, the motion of such a system occurs over the s-dimensional surface defined by Eq. (1). Generally speaking, the point representing the motion of the system can pass arbitrarily close to any point on the given surface. This becomes apparent when transitioning to the angle and action variables in the phase space. Figure 79 shows the projection of the phase trajectory onto the w_1-w_2 plane. Since q_i and p_i are periodic functions of w_i, it suffices to consider the change in the angular variables within $0 \le w_i \le 2\pi$. On the w_1-w_2 plane, the phase trajectory is located within a square, and its opposite sides must be identified (i.e. 'glued' to each other).

In the general case, if the ratio ω_1/ω_2 is irrational, then the phase trajectory densely fills the entire square. In the q-p space, the phase trajectory fills the 's-dimensional surface torus.' This type of motion is referred to as *conditionally periodic*.

If any of the relations ω_i/ω_j are rational numbers, then the projection of the phase trajectory onto the w_i-w_j plane becomes a closed curve, and the dimension of the domain filled by the trajectory in the phase space decreases.

In general cases, the variables in the Hamilton–Jacobi equation are not separable, and the phase trajectory fills the region of greater dimensions. For $\partial H/\partial t = 0$, only one integral of the equations of motion $H(q,p) = E$ is 'guaranteed', so this is the domain $2s - 1$ dimensions (we are not concerned with proving this fact in our course). Thus, the impossibility of separating variables in the general case is not due to our inability to find suitable curvilinear coordinates, but by the very nature of the system's motion.

Note that methods of successive approximations can be successfully applied for the motion of the systems that differ only slightly from the systems with separable variables, but just for small intervals of time. However, the problem becomes increasingly complicated with longer time intervals due to the occurrence of the so-called resonant phenomena.

Consider an example, in some respects opposite to the case of conditionally periodic motion. Suppose that a large number of balls move without friction on a flat, bounded table and collide absolutely elastically with each other and with the walls. It is hardly possible to observe such 'gas' anywhere other than on a display screen. However, it provides a good model of a real gas.

The laws of ball motion are very simple. Formally, the entire phase trajectory of a system of balls is determined by its initial point, i.e. the initial coordinates and momenta of the balls.

Let us see how the phase trajectory changes if the direction of one of the balls changes at the initial moment of time by a very small angle $\varphi_0 \sim 10^{-8}$. Such a motion is called a perturbed one. We will now investigate how the deflection angle changes at collisions. In this study, we confine ourselves to rough estimates, keeping in mind that the average

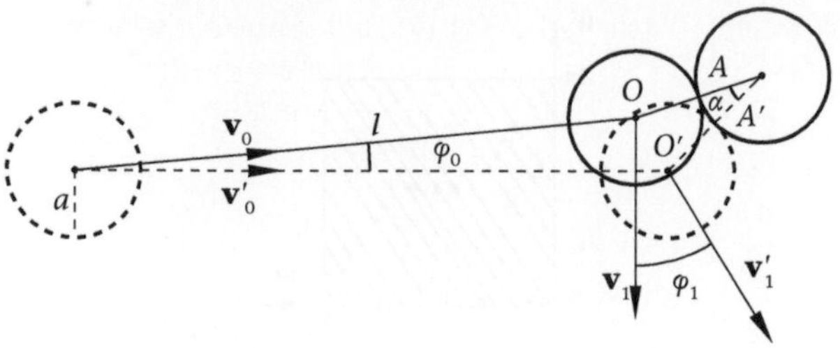

Figure 80 Elastic collision of the balls

distance travelled by the ball between collisions l is much greater than the ball radius a (i.e. $l \gg a$).

Before the next collision, the ball moves along a straight line, and its centre deviates from its unperturbed position by a distance $OO' \sim l\varphi_0$ (Fig. 80). The point of contact of the balls upon impact is offset by a similar amount:

$$AA' \sim \frac{1}{2}\,OO' \sim l\varphi_0, \quad \alpha \sim \frac{AA'}{a}, \quad \varphi_1 \sim \alpha \sim \frac{l}{a}\,\varphi_0.$$

The region of the surface of the second ball in the vicinity of the point of contact acts as a 'mirror' upon impact with our ball. The 'mirror' itself that has been moving before the impact bounces when hit. That is why the direction in which our ball bounces is not defined by the ordinary rule of 'the incident angle equals the reflection angle.' However, rotating the 'mirror' by a small angle α leads to the motion of the rebounding ball, changing direction by an angle $\varphi_1 \sim \alpha$, which, perhaps, is one and a half to two times more or less. Such a level of accuracy is perfectly suitable for this case.

As

$$\varphi_1 \sim \frac{l}{a}\,\varphi_0 \gg \varphi_0,$$

then upon impact, the deflection increases sharply. After k collisions, the deviation can be estimated as

$$\varphi_k \sim \left(\frac{l}{a}\right)^k \varphi_0.$$

This indicates that the perturbation of the deflection angle of the trial ball grows exponentially with time. With $l/a \sim 10$, it is sufficient to have $k \sim 8 \div 10$ collisions to obtain $\varphi_k \sim 1$. Then the perturbed direction of the ball motion has no relation to the unperturbed motion.

It turns out that the growth of perturbation is transferred to all balls, so the phase trajectory of the perturbed motion exponentially moves away from the phase trajectory of the unperturbed one. On the other hand, this divergence contributes to the establishment of the Maxwell distribution in velocity.

The difference in the behaviour of conditionally periodic systems and the system of colliding balls leads to a qualitative difference in the content of numerical simulations of these systems.

An example of a weakly perturbed conditionally periodic system is the solar system in which almost elliptical planet trajectories, perturbed by the attraction between them, can, nevertheless, be calculated with good accuracy for millennia, both forward and backward in time. This allows, for example, to 'retrodict' eclipses observed in antiquity.

A prediction of coordinates and velocities of colliding balls after some considerable time is fundamentally impossible by using a computer. The inevitable inaccuracies in setting initial data and rounding errors exponentially increase during calculations. Hence, aiming to anticipate the motion of the balls after 100 collisions would necessitate initial velocities set at a fantastic accuracy of ~ 100 signs.

Even imagining calculations conducted by a supercomputer handling infinite characters requires acknowledging that motion after a substantial (yet not very long) time interval is 'controlled' by increasingly distant signs following the decimal point in the initial conditions. In other words, from the usual point of view, this motion is random.

But at the same time, after just a dozen collisions, the Maxwell distribution perfectly describes distribution in velocity. With only several balls one can already see the occurrence of the 'molecular chaos', obtain distribution functions over velocity, and so on.

This conclusion about the catastrophic growth of uncertainties in coordinates also applies to the system of a large number of balls and the motion of molecules of a real gas, where uncertainties arise from all sorts of perturbations neglected otherwise (in comparison to our balls, where they arise due to limited calculation accuracy). The motion of the balls (and molecules) is relatively regular in small time intervals but random over an extended period. Note that this randomness is realized within the framework of the law of the energy conservation.

In conclusion, the origin of randomness in motion, strictly defined, remains a subject of intense modern physical exploration. Moreover, the notion that the phenomenon of the chaotic motion under seemingly complete certainty of the initial conditions is the rule rather than the exception, has already penetrated other fields of science and philosophy.

CHAPTER V

Rigid-body motion

Here, we define a *rigid body* as *a set of material points, whose distances do not change during motion*. This implies that we disregard all types of deformation, which, in many cases, is a reasonable approximation. However, it should be noted that this approximation is only valid in non-relativistic mechanics. The existence of an 'absolutely rigid body' contradicts the theory of relativity as it allows the transmission of signals at arbitrary velocities.

§ 52 Kinematics of a rigid body

The position of a rigid body in an inertial coordinate frame XYZ is determined as follows. Consider a coordinate system $x = x_1, y = x_2, z = x_3$ 'pinned' to a rigid body with the system's origin at a point O (Fig. 81). This coordinate system is called a 'moving' one. Then the position of the body is defined by six quantities, including three coordinates of the radius vector $\mathbf{R}$ for point O in the XYZ inertial frame and three angles that describe the orientation of the xyz axes relative to the XYZ axes.

If a rigid body consists of N material points with masses m_a and radius vectors $\mathbf{r}_a$ in the coordinate system xyz, then the radius vector of a point a in the inertial frame is

$$\mathbf{R} + \mathbf{r}_a.$$

Let us emphasize the fact, important for what follows, that *projections x_a, y_a, z_a of the $\mathbf{r}_a$ vector on the xyz axes of the moving coordinate system do not change when a rigid body moves*, although the vector $\mathbf{r}_a$ can change its direction (but not length!).

The arbitrary motion of a rigid body can be represented as a combination of two movements: translational and rotational. In *translational motion*, the vector $\mathbf{R}$ changes, but the vector $\mathbf{r}_a$ maintains its direction, so that all lines connecting pairs of points in the

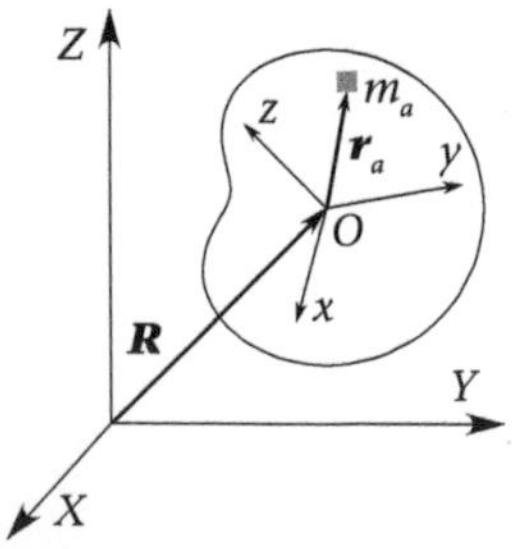

Figure 81 Inertial coordinate system XYZ and 'moving' coordinate system xyz 'pinned' to a rigid body

Lectures on Analytical Mechanics. G. L. Kotkin, V. G. Serbo, A. I. Chernykh, Oxford University Press. © G. L. Kotkin, V. G. Serbo, A. I. Chernykh (2024). DOI: 10.1093/oso/9780198894674.003.0005

rigid body remain parallel. Thus, the velocity $\mathbf{v}_a$ of any material point coincides with the velocity of the points of O, i.e.

$$\mathbf{v}_a = \dot{\mathbf{R}} \equiv \mathbf{V}\,. \tag{52.1}$$

In *rotational* motion, on the other hand, the position of the point O remains constant, and the rigid body rotates with angular velocity Ω around the direction given by the unit vector $\mathbf{n}$. It is convenient to introduce the vector of the angular velocity

$$\boldsymbol{\Omega} = \Omega \cdot \mathbf{n}\,.$$

This vector depends on time since the magnitude and direction of rotation may change. During the rotational motion of a rigid body, its constituent material points rotate with angular velocity $\boldsymbol{\Omega}$ in planes perpendicular to $\mathbf{n}$. Hence, the velocity of the a-th material point is (cf. § 20.2)

$$\mathbf{v}_a = [\boldsymbol{\Omega}, \mathbf{r}_a]\,. \tag{52.2}$$

The following formal consideration will illustrate this simple formula. We introduce the unit vectors $\mathbf{e}_i$ of the moving coordinate system $x_1x_2x_3$ with properties

$$\mathbf{e}_i\,\mathbf{e}_k = \delta_{ik}\,, \quad i,k = 1,\ 2,\ 3\,.$$

The vector $\mathbf{r}_a$ can be represented as an expansion in these orts:

$$\mathbf{r}_a = x_a\mathbf{e}_1 + y_a\mathbf{e}_2 + z_a\mathbf{e}_3\,. \tag{52.3}$$

When a rigid body rotates, the unit vectors of $\mathbf{e}_i$ also rotate with the same angular velocity $\boldsymbol{\Omega}$, so that

$$\frac{d\mathbf{e}_i}{dt} = [\boldsymbol{\Omega}, \mathbf{e}_i]\,. \tag{52.4}$$

At the same time, the coordinates $x_a,\ y_a,\ z_a$ of the vector $\mathbf{r}_a$, as noted above, do not change when a rigid body moves. Therefore, when differentiating relation (3) with respect to time, we obtain result (2).

For what follows, it is useful to consider also the time derivative for arbitrary time-dependent vector $\mathbf{A} = \mathbf{A}(t)$. Its expansion in the introduced orts

$$\mathbf{A}(t) = A_1(t)\mathbf{e}_1 + A_2(t)\mathbf{e}_2 + A_3(t)\mathbf{e}_3 \tag{52.5}$$

contains coordinates $A_1,\ A_2,\ A_3$, which, generally speaking, depend on time. Therefore, differentiation of relation (5) leads to another result:

$$\frac{d\mathbf{A}}{dt} = \frac{dA_1}{dt}\mathbf{e}_1 + \frac{dA_2}{dt}\mathbf{e}_2 + \frac{dA_3}{dt}\mathbf{e}_3 + [\boldsymbol{\Omega}, \mathbf{A}]\,. \tag{52.6a}$$

Projecting this equation onto the $\mathbf{e}_i$ axis, we obtain

$$\left(\frac{d\mathbf{A}}{dt}\right)_i \equiv \mathbf{e}_i \cdot \frac{d\mathbf{A}}{dt} = \frac{dA_i}{dt} + \sum_{j,k=1}^{3} e_{ijk}\,\Omega_j A_k\,, \quad i = 1,\ 2,\ 3\,. \tag{52.6b}$$

Here e_{ijk} is the complete antisymmetric unit tensor of the third rank.

When the rigid body moves arbitrarily, the velocity of the a-th material point is

$$\mathbf{v}_a = \mathbf{V} + [\boldsymbol{\Omega}, \mathbf{r}_a]\,. \tag{52.7}$$

The origin of the moving coordinate system xyz, i.e. the point O, may be chosen completely arbitrarily, including the position outside the rigid body. Let us show, however, that *the angular velocity* $\boldsymbol{\Omega}$ *does not depend on the choice of point* O. When the origin of the point O' is shifted by the vector $\mathbf{B}$ from the point O, then the new radius vector $\mathbf{r}'_a$ is related to the old radius vector $\mathbf{r}_a$ by equation

$$\mathbf{r}_a = \mathbf{r}'_a + \mathbf{B}\,. \tag{52.8}$$

In this case, the velocity of the point O' is (in accordance with (7))

$$\mathbf{V}' = \mathbf{V} + [\boldsymbol{\Omega}, \mathbf{B}]\,. \tag{52.9}$$

For the velocity $\mathbf{v}_a$ of a material point, we can use two equivalent expressions: either Eq. (7) in the old coordinates, or equation

$$\mathbf{v}_a = \mathbf{V}' + [\boldsymbol{\Omega}', \mathbf{r}'_a] \tag{52.10}$$

in new coordinates. Substituting (8) into Eq. (7), we obtain relation

$$\mathbf{V}' + [\boldsymbol{\Omega}', \mathbf{r}'_a] = \mathbf{V} + [\boldsymbol{\Omega}, \mathbf{B}] + [\boldsymbol{\Omega}, \mathbf{r}'_a]\,,$$

from which, by virtue of (9) and the arbitrariness of the vector $\mathbf{r}'_a$, the equality follows

$$\boldsymbol{\Omega}' = \boldsymbol{\Omega}\,,$$

i.e. independence of the angular velocity of rotation of a rigid body from position of the point O.

The velocity $\mathbf{V}$ can be decomposed into components $\mathbf{V}_\parallel$ and $\mathbf{V}_\perp$ parallel and perpendicular to the vector of the angular velocity $\boldsymbol{\Omega}(t)$ in a given time. Let us do the same for the velocity $\mathbf{V}'$, then from the equality (9), we obtain two relations:

$$\mathbf{V}'_\parallel = \mathbf{V}_\parallel\,, \quad \mathbf{V}'_\perp = \mathbf{V}_\perp + [\boldsymbol{\Omega}, \mathbf{B}]\,.$$

If $\mathbf{V}_\parallel = 0$, then there exists a point O' such that $\mathbf{V}' = 0$ at this moment of time, and the motion of the rigid body can be expressed as a pure rotation. The corresponding axis of rotation passing through the point O' is called *an instantaneous axis.*

§ 53 Momentum, angular momentum and kinetic energy of a rigid body

53.1 *Momentum of a rigid body*

The momentum of a rigid body is given by the sum of the momenta of its constituent material points, which can be expressed as follows:

$$\mathbf{P} = \sum_{a=1}^{N} m_a \mathbf{v}_a\,. \tag{53.1a}$$

Substituting (52.7) and introducing the notation for the mass of the entire rigid body and radius vector of its centre of mass

$$m = \sum_a m_a\,, \quad \mathbf{r}_{\rm cm} = \frac{\sum_a m_a \mathbf{r}_a}{m}\,, \tag{53.2}$$

we find

$$\mathbf{P} = m\mathbf{V} + m\,[\boldsymbol{\Omega}, \mathbf{r}_{\rm cm}]\ . \tag{53.1b}$$

If the origin of the moving coordinate system xyz is placed at the centre of mass, such that $\mathbf{r}_{\rm cm} = 0$, *the momentum of the rigid body will be reduced to that of a material point with mass mand velocity equal to the velocity of the centre of mass* $\mathbf{V}_{\rm cm}$:

$$\mathbf{P} = m\mathbf{V}_{\rm cm}. \tag{53.3}$$

53.2 *Angular moment of a rigid body*

The angular momentum of a rigid body is equal to the sum of the angular momenta of its constituent material points:

$$\mathbf{M} = \sum_{a=1}^{N} m_a\,[\mathbf{R} + \mathbf{r}_a,\ \mathbf{V} + [\boldsymbol{\Omega}, \mathbf{r}_a]] = m[\mathbf{R}, \mathbf{V}] +$$

$$+ \sum_a m_a\,[\mathbf{r}_a,\ [\boldsymbol{\Omega}, \mathbf{r}_a]] + m[\mathbf{r}_{\rm cm}, \mathbf{V}] + m\,[\mathbf{R}, [\boldsymbol{\Omega}, \mathbf{r}_{\rm cm}]]\ . \tag{53.4}$$

In the case when the origin of the coordinate system xyz is at the centre of mass, *the angular momentum can be expressed as the sum of the angular momentum of a material point with mass m, radius vector* $\mathbf{R}$*, and velocity equal to the velocity of motion of the centre of mass, and the angular momentum corresponding to the rotation of the rigid body about the point O with angular velocity* $\boldsymbol{\Omega}$:

$$\mathbf{M} = m[\mathbf{R}, \mathbf{V}_{\rm cm}] + \sum_a m_a\,[\mathbf{r}_a,\ [\boldsymbol{\Omega}, \mathbf{r}_a]]\ . \tag{53.5}$$

When the point O is at rest, the angular momentum of the rigid body is expressed by

$$\mathbf{M} = \sum_a m_a\,[\mathbf{r}_a,\ [\boldsymbol{\Omega}, \mathbf{r}_a]]\ . \tag{53.6}$$

In this case, the components of the angular momentum can be represented as a linear form of the components of the angular velocity Ω_j. Indeed, rewriting

$$[\mathbf{r}_a, [\boldsymbol{\Omega}, \mathbf{r}_a]] = \boldsymbol{\Omega}\,\mathbf{r}_a^2 - (\boldsymbol{\Omega}\,\mathbf{r}_a)\,\mathbf{r}_a\,,$$

and projecting this equality onto the axes of the moving coordinate system $x_1x_2x_3$,

$$[\mathbf{r}_a, [\boldsymbol{\Omega}, \mathbf{r}_a]] \cdot \mathbf{e}_i = \sum_{k=1}^{3} \left\{\mathbf{r}_a^2\,\delta_{ik} - (\mathbf{r}_a)_i\,(\mathbf{r}_a)_k\right\}\,\Omega_k\,,$$

we represent the components of the angular momentum in the form

$$M_i = \sum_{k=1}^{3} I_{ik}\,\Omega_k\,, \tag{53.7}$$

with the coefficients I_{ik} corresponding to the symmetric tensor of the second rank—the so-called *inertia tensor* of the rigid body:

$$I_{ik} = \sum_{a=1}^{N} m_a \left\{\mathbf{r}_a^2\,\delta_{ik} - (\mathbf{r}_a)_i\,(\mathbf{r}_a)_k\right\}\,, \quad i,\,k = 1,\,2,\,3\,. \tag{53.8}$$

We want to emphasize that the angular velocity and angular momentum, generally speaking, depend on time, while the components of I_{ik} are constant characteristics of a rigid body. Properties of this tensor will be described in § 53.4. Eq. (7) shows that, in the general case, the direction of the vector **M** and the vector **Ω** do not coincide.

53.3 *Kinetic energy of a rigid body*

The expression for the kinetic energy of a rigid body is given by the summate of the kinetic energies of its constituent material points:

$$T = \frac{1}{2}\sum_{a=1}^{N} m_a \mathbf{v}_a^2\,.$$

Using the notation (2) and substituting (52.7) into the above equation, we can simplify it as follows

$$T = \frac{1}{2}\,m\mathbf{V}^2 + \frac{1}{2}\sum_a m_a[\mathbf{\Omega},\mathbf{r}_a]^2 + m\,\mathbf{\Omega}[\mathbf{r}_{\rm cm},\mathbf{V}]\,.$$

When the origin of the moving coordinate system *xyz* is at the centre of mass of the rigid body, *the kinetic energy of the rigid body is the sum of the kinetic energy of a material point with mass m and velocity* $\mathbf{V}_{\rm cm}$ *and the kinetic energy of rotation of the rigid body*:

$$T = \frac{1}{2}\,m\mathbf{V}_{\rm cm}^2 + \frac{1}{2}\sum_a m_a[\mathbf{\Omega},\mathbf{r}_a]^2\,. \tag{53.9}$$

When the point *O* is at rest, i.e. all the kinetic energy of the rigid body is the energy of rotation, this kinetic energy can be represented as follows:

$$T = \frac{1}{2}\sum_a m_a[\mathbf{\Omega},\mathbf{r}_a]^2\,. \tag{53.10}$$

It may be useful to compare kinetic energies of rotation and translational motion of a rigid body. The latter can be represented as a scalar product of the momentum of the body and its velocity:

$$T = \frac{1}{2}\,m\mathbf{V}^2 = \frac{1}{2}\,\mathbf{P}\mathbf{V}\,. \tag{53.11}$$

Similarly, we can represent the formula (1) for the kinetic energy of rotation as a scalar product of the angular momentum vector of the body and the vector of its angular velocity:[1]

$$T = \frac{1}{2}\mathbf{M}\boldsymbol{\Omega}\,. \tag{53.12}$$

To do this, it suffices to use Eq. (6) and rewrite

$$(\boldsymbol{\Omega}\,[\mathbf{r}_a,\,[\boldsymbol{\Omega},\mathbf{r}_a]]) = [\boldsymbol{\Omega},\mathbf{r}_a]^2\,.$$

By substituting the expression for the angular momentum in the form (7) into equation (12), we obtain the kinetic energy of the rigid body as a quadratic form of the components of the angular velocity Ω_j:

$$T = \frac{1}{2}\sum_{i,k=1}^{3} I_{ik}\,\Omega_i\Omega_k\,, \tag{53.13}$$

where the coefficients of the quadratic form I_{ik} are the components of the inertia tensor of a rigid body (8).

53.4 *Inertia tensor of a rigid body*

Both the angular momentum of a rotating rigid body and its kinetic energy can be expressed in terms of the inertia tensor. Regardless of their mass distribution, two bodies with the same inertia tensor will rotate the same way. Let us consider the properties of the inertia tensor in more detail.

The diagonal elements of the inertia tensor are calculated using the squared distances between material points and the corresponding axis:

$$I_{11} = \sum_a m_a(y_a^2 + z_a^2), I_{22} = \sum_a m_a(x_a^2 + z_a^2), I_{33} = \sum_a m_a(x_a^2 + y_a^2)\,.$$

On the other hand, the off-diagonal elements are determined through the products of the corresponding coordinates:

$$I_{12} = I_{21} = -\sum_a m_a x_a y_a\,,\;\; I_{13} = I_{31} = -\sum_a m_a x_a z_a\,,\;\; I_{23} = I_{32} = -\sum_a m_a y_a z_a\,.$$

It can be seen from these definitions that the sum of two different diagonal elements of the inertia tensor I_{ik} can be no less than the third diagonal element, that is,

$$I_{11} + I_{22} = \sum_a m_a(x_a + y_a^2 + 2z_a^2) \geq \sum_a m_a(x_a^2 + y_a^2) = I_{33}\,.$$

From the same relation it can be seen that for a flat rigid body (located in the xy-plane)

$$I_{11} + I_{22} = I_{33}\,.$$

[1] Note that since $T > 0$, the angle between vectors $\mathbf{M}$ and $\boldsymbol{\Omega}$ is always less than 90°.

Let the unit vector $\mathbf{n}$ define the direction of some axis, and the value ρ_a denote the distance from the material point a to this axis. Determine the *moment of inertia* of a rigid body with respect of the $\mathbf{n}$ axis as follows

$$I_{\mathbf{n}} = \sum_a m_a \rho_a^2 .$$

The diagonal elements of the inertia tensor, I_{11}, I_{22}, and I_{33}, represent the moments of inertia with respect to the x, y, and z axes, respectively. On the other hand, the moment of inertia for an arbitrary axis $\mathbf{n}$ can be expressed in terms of the components of tensor I_{ik} and vector $\mathbf{n}$ as:

$$I_{\mathbf{n}} = \sum_{i,k=1}^{3} I_{ik}\, n_i n_k .$$

Consider the case when the origin O of the moving coordinate system xyz is placed at the centre of mass of the rigid body, so that

$$\sum_a m_a \mathbf{r}_a = 0 , \tag{53.14}$$

and the point O' (the origin of another moving coordinate system $x'y'z'$) has shifted by the vector $\mathbf{B}$, whereby the radius vectors $\mathbf{r}_a$ and $\mathbf{r}'_a$ of a material point satisfy relation (52.8). By substituting this relation into definition (8) and taking into account equality (14), the following useful formula (sometimes called the *Huygens–Steiner theorem*) is obtained:

$$I'_{ik} = I_{ik}^{\mathrm{cm}} + m\left(\mathbf{B}^2 \delta_{ik} - B_i B_k\right) . \tag{53.15}$$

It is evident from this equation that the diagonal elements of the inertia tensor have their smallest value at the centre of mass of the rigid body.

Similar to any other symmetric tensor, the inertia tensor may also be diagonalized with a suitable choice the axes of the moving coordinate system xyz relative to the rigid body. A special significance is held by the diagonalization of this tensor in the case when the origin of the system xyz is placed at the centre of mass of a rigid body. The diagonal elements of the inertia tensor I_{ik}^{cm} in this case, which are

$$I_1 \equiv I_{11}^{\mathrm{cm}} , \quad I_2 \equiv I_{22}^{\mathrm{cm}} , \quad I_3 \equiv I_{33}^{\mathrm{cm}} , \tag{53.16}$$

are referred to as the *principal moments of inertia* of the rigid body, while the respective axes xyz are called the *principal axes of inertia*.

Definitions *If the principal moments of inertia of a rigid body are equal, $I_1 = I_2 = I_3$, the body is referred to as a* spherical top. *A homogeneous ball with radius R is an example of a spherical top with*

$$I_1 = I_2 = I_3 = \frac{2}{5} mR^2 . \tag{53.17}$$

However, a homogeneous cube of side a is also a spherical top, for which

$$I_1 = I_2 = I_3 = \frac{1}{6} ma^2 . \tag{53.18}$$

If a rigid body has the same two principal moments of inertia, for example $I_1 = I_2 \neq I_3$, then such a body is called a symmetrical top. *One example of that is a homogeneous ellipsoid of revolution with semi-axes $a = b \neq c$, for which*

$$I_1 = I_2 = \frac{1}{5}\, m\,(a^2 + c^2), \;\; I_3 = \frac{2}{5}\, ma^2\,. \tag{53.19}$$

Another example of a symmetrical top is four points with equal masses $m_1 = m_2 = m_3 = m_4$ located at the vertices of a square and connected by weightless rods of length a (see Fig. 91 for $M = m$), in this case

$$I_1 = I_2 = m_1 a^2, \;\; I_3 = 2\, m_1 a^2\,. \tag{53.20}$$

If a rigid body has different principal moments of inertia, then the body is called an asymmetrical top.

Problems

53.1. Particles of mass m and M are located at the vertices of a square of side $2a$ (Fig. 82). Find components of inertia tensor with respect to:
a) axes xyz;
b) axes $x'y'$ coinciding with the diagonals of the square, and z.

53.2. Find the principal axes of inertia and principal moments of inertia of a system in which particles of mass m and $2m$ are located at the vertices of a right triangle with legs $2a$ and $4a$ (Fig. 83).

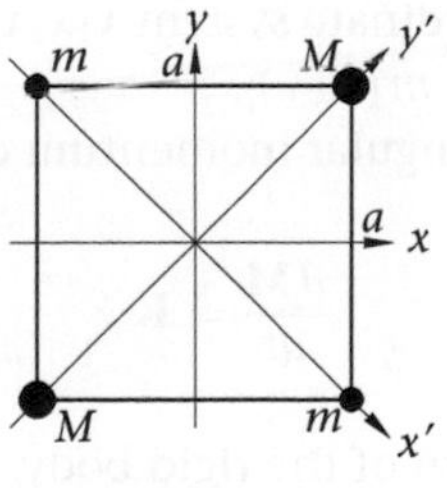

Figure 82 Four particles at the vertices of a square

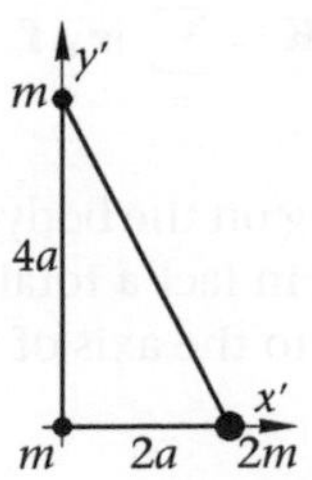

Figure 83 Particles at the vertices of a triangle

§ 54 Equations of motion of a rigid body. Examples

54.1 *Equations of motion of a rigid body. Euler equations*

The equation for the momentum of a rigid body can be derived using Newton's law for the motion of the a-th material point under the force $\mathbf{f}_a$:

$$\frac{d(m_a \mathbf{v}_a)}{dt} = \mathbf{f}_a \,.$$

Summing these equations over all material points of a rigid body, we get

$$\frac{d\mathbf{P}}{dt} = \mathbf{F}\,, \tag{54.1}$$

where $\mathbf{P}$ is the momentum of the rigid body defined by the formulae (52.1)–(52.3), and

$$\mathbf{F} = \sum_a \mathbf{f}_a$$

is the total force acting on the body. Internal forces acting between the material points of a rigid body are mutually compensated due to the Newton's third law. Therefore, the force $\mathbf{F}$ is in fact the total external force acting on the rigid body. Projecting equations (1) onto the axes of the moving coordinate system $x_1x_2x_3$ and using (52.6), we obtain

$$\frac{dP_i}{dt} + \sum_{j,k=1}^{3} e_{ijk}\, \Omega_j P_k = F_i\,, \quad i = 1,\, 2,\, 3\,. \tag{54.2}$$

If the origin of the moving coordinate system $x_1x_2x_3$ is placed at the centre of mass of a rigid body (see (53.3)), then $P_i = m\,(\mathbf{V}_{\rm cm})_i$.

Similarly, the equation for the angular momentum of a rigid body can be obtained:

$$\frac{d\mathbf{M}}{dt} = \mathbf{K}\,, \tag{54.3}$$

where $\mathbf{M}$ is the angular momentum of the rigid body, defined by formulae (53.4)–(53.7), and

$$\mathbf{K} = \sum_a \,[\mathbf{r}_a, \mathbf{f}_a]$$

is the total momentum of forces acting on the body. Since the internal moments of forces are mutually compensated, then $\mathbf{K}$ is in fact a total moment of external forces acting on the body. Projecting equations (3) onto the axis of the moving coordinate system $x_1x_2x_3$ and using (52.6), we obtain

$$\frac{dM_i}{dt} + \sum_{j,k=1}^{3} e_{ijk}\, \Omega_j M_k = K_i\,, \quad i = 1,\, 2,\, 3\,. \tag{54.4}$$

If the origin of the moving coordinate system $x_1x_2x_3$ is at rest, then (see (53.7))

$$\frac{dM_i}{dt} = \sum_{k=1}^{3} I_{ik}\frac{d\Omega_k}{dt}, \quad i = 1, 2, 3. \tag{54.5}$$

If, additionally, the axes of system $x_1x_2x_3$ are chosen along the principal axes of inertia, then Eqs. (4), called *the Euler equations*, look especially simple.

$$I_i\frac{d\Omega_i}{dt} + \sum_{j,k=1}^{3} e_{ijk}\,\Omega_j\Omega_k\,I_k = K_i\,, \quad i = 1, 2, 3. \tag{54.6}$$

Several examples of the application of Eqs. (3)–(6) are considered below.

54.2 *Free motion of spherical and symmetrical tops*

Let us consider the case of free motion of a rigid body. Since $\mathbf{F} = \mathbf{K} = 0$, the momentum and angular momentum of the body are conserved, and the centre of mass moves with a constant velocity. An inertial frame of reference can be chosen such that the centre of mass is at rest, and the origin of the inertial frame XYZ and the moving frame xyz are intersecting at the centre of mass of the rigid body.

The spherical top has $I_{ik} = I\,\delta_{ik}$, so from (53.7) it follows that

$$\mathbf{M} = I\boldsymbol{\Omega}. \tag{54.7}$$

From this it is seen that during the free motion of a spherical top, not only the angular momentum $\mathbf{M}$, but also the angular velocity vector $\boldsymbol{\Omega}$ is conserved. Moreover, the vectors $\mathbf{M}$ and $\boldsymbol{\Omega}$ have the same direction.

The laws of conservation of momentum and angular momentum are also sufficient for definition of the free motion of a symmetrical top. If the x_3-axis (its unit vector $\mathbf{e}_3$) coincides with the axis of symmetry of the top, then $I_1 = I_2 \neq I_3$, and

$$M_1 = I_1\,\Omega_1\,, \;\; M_2 = I_1\,\Omega_2\,, \;\; M_3 = I_3\,\Omega_3\,. \tag{54.8}$$

In vector form, these equations can be represented as

$$\boldsymbol{\Omega} = \frac{1}{I_1}\,(M_1\mathbf{e}_1 + M_2\mathbf{e}_2) + \frac{M_3}{I_3}\mathbf{e}_3\,,$$

or in the form

$$\boldsymbol{\Omega} = \frac{\mathbf{M}}{I_1} + \left(\frac{M_3}{I_3} - \frac{M_3}{I_1}\right)\mathbf{e}_3\,. \tag{54.9}$$

This formula is very convenient in analysing the motion of a top. It shows that the three vectors $\mathbf{M}$, $\boldsymbol{\Omega}$, and $\mathbf{e}_3$ always reside in a common plane. In the inertial coordinate system

XYZ, the vector **M** is motionless, and the motion of the unit vector $\mathbf{e}_3$ is determined by Eq. (52.4). If we take into account (9), this equation now reads

$$\frac{d\mathbf{e}_3}{dt} = [\boldsymbol{\Omega}, \mathbf{e}_3] = \left[\frac{\mathbf{M}}{I_1}, \mathbf{e}_3\right] . \qquad (54.10)$$

Therefore, the axis of symmetry of the top rotates about the direction of angular momentum at a constant angular velocity of $\mathbf{M}/I_1$. Under such rotation, the projection M_3 of the angular momentum on the symmetry axis remains unchanged, and the angular velocity vector $\boldsymbol{\Omega}$ rotates with the same angular velocity as the symmetry axis.

In this way, the three vectors $\mathbf{M}$, $\boldsymbol{\Omega}$ and $\mathbf{e}_3$ remain in the same plane and retain their mutual arrangement and lengths. Thus, in the considered inertial coordinate system (in which angular momentum is conserved), the vectors $\boldsymbol{\Omega}$ and $\mathbf{e}_3$ lie in the same plane and rotate along conical surfaces around the direction of angular momentum (see Fig. 84) with the same angular velocity. This type of motion is called a *regular precession*, whose angular velocity is known as the *precession velocity* $\boldsymbol{\Omega}_{\text{pr}}$, given by

$$\boldsymbol{\Omega}_{\text{pr}} = \frac{\mathbf{M}}{I_1} . \qquad (54.11)$$

In addition to regular precession, there exists another motion of the top known as *proper rotation*. This type of motion involves the rotation of the top about a rotating axis of symmetry. The corresponding angular velocity of the proper rotation is given by

$$\boldsymbol{\Omega}_{\text{prop. rot.}} = \left(\frac{M_3}{I_3} - \frac{M_3}{I_1}\right) \mathbf{e}_3 = \left(1 - \frac{I_3}{I_1}\right) \Omega_3\, \mathbf{e}_3 . \qquad (54.12)$$

The angular velocities of these two rotations (the precessions $\boldsymbol{\Omega}_{\text{pr}}$ and proper rotation $\boldsymbol{\Omega}_{\text{prop. rot.}}$) in the sum is the total angular velocity of the spinning top:

$$\boldsymbol{\Omega} = \boldsymbol{\Omega}_{\text{pr}} + \boldsymbol{\Omega}_{\text{prop. rot.}} . \qquad (54.13)$$

A visual interpretation of proper rotation can be obtained by transitioning from the considered inertial coordinate system to the coordinate system that is rotating with an angular velocity $\boldsymbol{\Omega}_{\text{pr}}$. In this system, the top rotates with an angular velocity that is equal to the difference $\boldsymbol{\Omega} - \boldsymbol{\Omega}_{\text{pr}}$ and precisely equals the angular velocity of the proper rotation, $\boldsymbol{\Omega}_{\text{prop. rot.}}$.

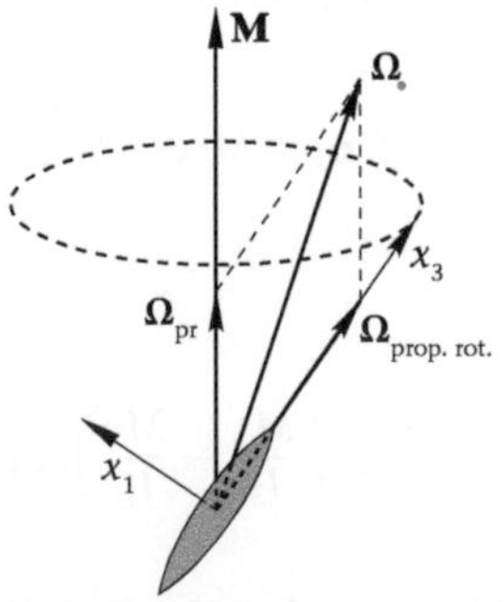

Figure 84 Free motion of a symmetrical top

As an illustration of the results obtained, we consider the free motion of the *matryoshka* in the inertial coordinate system described above. Usually, a matryoshka is a body of revolution with the x_3 axis as the symmetry axis, passing through the fixed centre of mass of the matryoshka, and rotating with an angular velocity $\boldsymbol{\Omega}_{\rm pr}$ around the direction of the angular momentum. For illustrative purposes, we assume that at the initial moment of time, the angle between the angular momentum $\mathbf{M}$ and the symmetry axis equals $\alpha_M = 60^\circ$. We can then find the angle α_Ω between the angular velocity $\boldsymbol{\Omega}$ and the symmetry axis from the equation:

$$\tan\alpha_\Omega = \frac{\Omega_\perp}{\Omega_3} = \frac{M_\perp/I_1}{M_3/I_3} = \frac{I_3}{I_1}\tan\alpha_M\,, \tag{54.14}$$

$$\mathbf{M}_\perp \equiv M_1\mathbf{e}_1 + M_2\mathbf{e}_2 = I_1\boldsymbol{\Omega}_\perp\,.$$

During the subsequent motion of the matryoshka, both of these angles will retain their values. If the matryoshka is a monochrome body of revolution, we will not be able to observe its proper rotation. For example, after a time $T_{\rm pr} = 2\pi/\Omega_{\rm pr}$, the symmetry axis of the matryoshka will return to its initial position, and it will be impossible to distinguish the new position of the matryoshka from the initial one. However, if the matryoshka is painted (and a face is drawn on it), and the matryoshka has its face in a full-face position in the initial moment of time, after a time $T_{\rm pr}$, we will observe this face turning at an angle:

$$\psi = \Omega_{\text{prop. rot.}}\cdot T_{\rm pr} = 2\pi\frac{\Omega_{\text{prop. rot.}}}{\Omega_{\rm pr}} = 2\pi\left(\frac{I_1}{I_3}-1\right)\cos\alpha_M\,. \tag{54.15}$$

Consider these two options: a 'thin' matryoshka,

$$I_1 = 2I_3\,,$$

and 'full' matryoshka,

$$I_1 = \frac{3}{34}I_3\,.$$

In the first case, the angle $\alpha_\Omega \approx 40^\circ$, the location of vectors $\mathbf{M}$, $\boldsymbol{\Omega}$ and $\mathbf{e}_3$ are approximately the same as in Fig. 84, and after a time $T_{\rm pr}$, the angle $\psi = \pi$ (that is, a 'thin' matryoshka turns its back to the observer). Try to define yourself the mutual arrangement of vectors $\mathbf{M}$, $\boldsymbol{\Omega}$ and $\mathbf{e}_3$ for 'full' matryoshka and its rotation angle ψ in time $T_{\rm pr}$.

Finally, we consider the motion of a symmetrical top, which, in the inertial coordinate system, has a fixed point O located on the axis of symmetry. Let $\mathbf{l}$ be the vector drawn from point O to the centre of mass of the top along the symmetry axis. If the forces acting on the top are applied only at a fixed point O which is the origin of the inertial frame of coordinates XYZ, the moment of these forces equals zero. Therefore, the angular momentum of the top is conserved. It is easy to see that in this case, the equations of motion of a free top (8)–(13) retain their form when replacing

$$I_1 \to I_1' = I_1 + ml^2\,. \tag{54.16}$$

In particular, the symmetry axis of the top and the vector **l** rotate with the same angular velocity

$$\mathbf{\Omega}_{\rm pr} = \frac{\mathbf{M}}{I_1'} \tag{54.17}$$

around a constant vector of the angular momentum **M**.

54.3 *Fast top in the field of gravity*

In this section, we consider the motion of the top described in the previous section when, in addition to the force applied at a stationary point O, the force of gravity $m\mathbf{g}$ is also acting. To illustrate, we assume that the fixed point is located below the centre of mass of the top, as shown in Fig. 85. The general solution to this problem can be found in [1], § 35, problem 1. We will only consider the case of a 'fast top' with large kinetic energy

$$T \sim \frac{M^2}{I_3} \gg mgl\,. \tag{54.18}$$

Under these conditions, we can neglect the influence of the gravitational force in the first approximation. Consequently, we arrive at the same problem as described in the previous subsection, with constant angular momentum **M** and a symmetry axis precessing around **M** with an angular velocity (17). In the current problem, this velocity is called the *angular velocity of nutation* and determined by the following expression

$$\mathbf{\Omega}_{\rm nut} = \frac{\mathbf{M}}{I_1'}\,. \tag{54.19}$$

where I_1' is the moment of inertia of the top about an axis perpendicular to the symmetry axis passing through the centre of mass.

In the next approximation, we take into account the effect of gravity, so the equation for the angular momentum of the top (3) takes the form

$$\frac{d\mathbf{M}}{dt} = [\mathbf{l},\, m\mathbf{g}]\,. \tag{54.20}$$

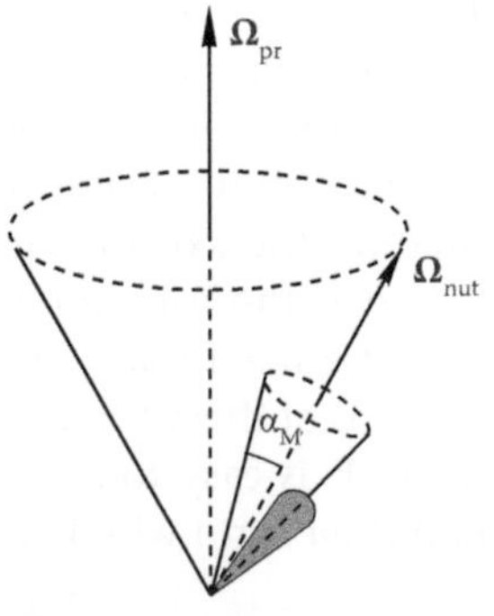

Figure 85 Fast top in the field of gravity

An approximate solution to this equation can be obtained by separating the fast and slow motions of the vectors $\mathbf{M}$ and $\mathbf{l}$, as done in § 37 for the Kapitsa's pendulum problem. Specifically, we represent these vectors as $\mathbf{M} = \langle\mathbf{M}\rangle + \delta\mathbf{M}$ and $\mathbf{l} = \langle\mathbf{l}\rangle + \delta\mathbf{l}$, where $\langle\ldots\rangle$ means averaging over fast rotation with angular velocity (19).

Further, we find that

$$\langle\mathbf{l}\rangle = \frac{\mathbf{M}}{M}\, l\cos\alpha_M\,, \tag{54.21}$$

where α_M is the angle between $\mathbf{M}$ and the symmetry axis of the top. To describe slowly changing average values, Eq. (20) is used which yields

$$\frac{d\langle\mathbf{M}\rangle}{dt} = [\langle\mathbf{l}\rangle,\, m\mathbf{g}] = [\boldsymbol{\Omega}_{\mathrm{pr}},\, \langle\mathbf{M}\rangle]\,, \quad \boldsymbol{\Omega}_{\mathrm{pr}} = -\frac{ml\cos\alpha_M}{M}\,\mathbf{g}\,. \tag{54.22}$$

This shows that the vector of the angular momentum precesses (rotates with angular velocity Ω_{pr}) around the vertical direction. Due to inequality (18), the ratio of the velocities Ω_{pr} and Ω_{nut} (19), as expected, turns out to be small:

$$\frac{\Omega_{\mathrm{pr}}}{\Omega_{\mathrm{nut}}} \sim \frac{mgl}{M^2/I_1'} \sim \frac{mgl}{T} \ll 1\,. \tag{54.23}$$

As a result, the symmetry axis of a fast top rotates with a large angular velocity (19) around the direction of the angular momentum, while the angular momentum itself slowly rotates around the vertical direction with angular velocity Ω_{pr} on average.

Here is another example. The Earth may be regarded as a slightly oblate ellipsoid, i.e. a fast symmetrical spinning top (not a spherical ball!). The attractions of the Sun and the Moon cause precession of the Earth's axis with a period of about 26 thousand years (the so-called *precession of the equinoxes*, see [3], problem 9.17).

Problems

54.1. Two identical uniform balls rotating with equal angular velocity ω slowly approach each other closely and then rigidly connect with each other. Determine the motion of the newly formed body and find which part of the initial kinetic energy has been converted into heat. Prior to connecting, the angular velocities of the balls were directed:

a) perpendicular to the line of the balls' centres and parallel to each other;

b) one along the line of the balls' centres, and the other perpendicular to this line.

54.2. A top with a fixed fulcrum O touches the horizontal plane with the edge of its disc (see Fig. 86). Before the touchdown, the top was spinning about its axis with an

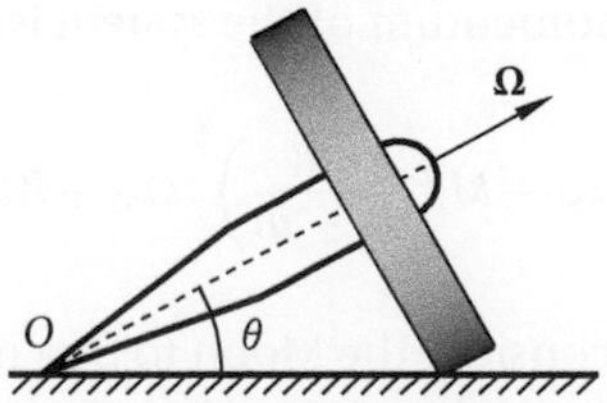

Figure 86 Spinning top with a fixed fulcrum O

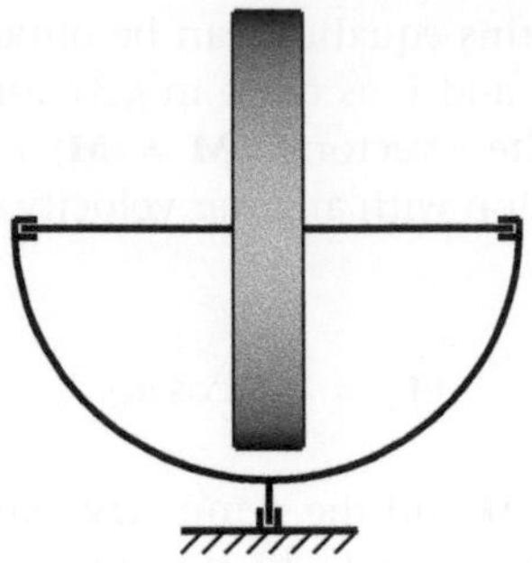

Figure 87 Gyrocompass

angular velocity Ω (assume the velocity of precession to be small). Find the angular velocity of the top when slipping of the disc vanishes. There were no nutations at the time of the touchdown.

54.3. A gyrocompass is a rapidly rotating disk whose spanning axis is confined to a horizontal plane (see Fig. 87). Study the motion of the gyrocompass at latitude λ. Angular velocity of the Earth's rotation equals ω.

§ 55 Effect of tidal forces on the length of a day and month

The modern period of the Earth's rotation around its axis $T_E = 2\pi/\Omega_E$ is approximately 28 times less than the period of rotation of the Moon around the Earth $T_M = 2\pi/\Omega_M$ (known as the month). Therefore, the ratio of the corresponding angular velocities can be approximately given by $\Omega_E/\Omega_M \approx 28$. It is well established that tidal forces increase the period of the Earth's rotation around its axis. Thus, it is interesting to determine the period of Earth's rotation at the moment it becomes equal to one month.

To simplify the calculation, the Earth's axis of rotation is assumed to be perpendicular to the plane of the Earth and Moon orbits. For numerical estimates, we consider the Earth as a uniform ball with a radius of $a = 6.4$ thousand km and a mass of M, which is 81 times greater than the mass of the Moon m. The distance from the Earth to the Moon is $R = 380$ thousand km.

The modern distances from the centre of mass of the Earth–Moon system to the Earth and to the Moon are equal to

$$\frac{m}{M+m}R \text{ and } \frac{M}{M+m}R,$$

respectively, and the angular momentum of the system is equal to

$$m\left(\frac{MR}{M+m}\right)^2 \Omega_M + M\left(\frac{mR}{M+m}\right)^2 \Omega_M + I\Omega_E = J\Omega_M + I\Omega_E, \tag{55.1}$$

where $J = MmR^2/(M+m)$. We consider the Moon to be a material point while taking into account the Earth's rotation around the centre of mass and around its own axis (with the moment of inertia $I = \frac{2}{5}Ma^2$).

At the instant when the period of Earth's rotation around its axis becomes equal to a month, the angular velocity of the Earth rotation ω equates to the angular velocity of the Moon's rotation, and the distance from the Earth to the Moon (according to Kepler's third law) attains $R(\Omega_M/\omega)^{2/3}$. The angular momentum can then be represented as

$$\left[J\,(\Omega_M/\omega)^{4/3}+I\right]\omega. \tag{55.2}$$

From Eqs. (1) and (2) we find the equation for $x=\omega/\Omega_E$:

$$x(1+k-x)^3=k^3\,\frac{\Omega_M}{\Omega_E}, \tag{55.3}$$

where

$$k=\frac{J\Omega_M}{I\Omega_E}=\frac{5}{2}\left(\frac{R}{a}\right)^2\frac{m}{M+m}\frac{\Omega_M}{\Omega_E}\approx 3.8.$$

Equation (3) (or in approximate form $x(4.8-x)^3=2.01$) has two real root: $x_1\approx 1/55$, corresponding to the future, and $x_2\approx 4$, corresponding to the past. Therefore, in the first case, it will take 55 modern days for the month to become equal to the Earth's rotation period, and in the second case, it was supposed to be 6 hours in this calculational model. Accordingly, the distance from the Earth to the Moon will become equal to $\approx 1.6R$, whereas it was equal to $2.6\,a$ (in this model!).

For more realistic models of the Earth–Moon system's evolution than considered here, see, for example, Chapter 2 in [17].

§ 56 Euler angles

Orientation of the moving coordinate system $x_1x_2x_3$ associated with a rigid body, denoted below as K, with respect to the fixed system XYZ (hereinafter – K_0) is customarily specified using three angles,[2] called the *Euler angles*.

Assume that the moving coordinate system K coincides with the system K_0 at the beginning. The transition to the final K system can be done with the help of three successive turns. The first rotation is by the angle φ around the Z-axis. The axis x_1 of the resulting coordinate system is called the *line of nodes N*, and it coincides with the line intersections of the old plane XY and the new plane x_1x_2. The second turn is a rotation through the angle θ around the line of nodes.[3] Let us take the third turn around the x_3-axis of the resulting coordinate system on the angle ψ. The Euler angles φ, θ, and ψ are called respectively *angles of precession, nutation, and proper rotation.* The angle θ takes values from 0 to π, and φ and ψ from 0 to 2π. The corresponding vectors of the angular velocities $\dot{\boldsymbol{\varphi}}$, $\dot{\boldsymbol{\theta}}$, and $\dot{\boldsymbol{\psi}}$ are directed along the Z axis, along the line of nodes N, and along the x_3 axis, respectively (Fig. 88).

[2] If, for example, we are talking about a bicycle wheel, then this could be the angle that defines the direction of motion, the angle of inclination of the wheel plane, and the angle fixed by the revolution counter.

[3] If we chose the axis for the second rotation x_2, then θ and φ would be spherical coordinates of the x_3-axis. This choice is preferred in the quantum theory of the angular momentum.

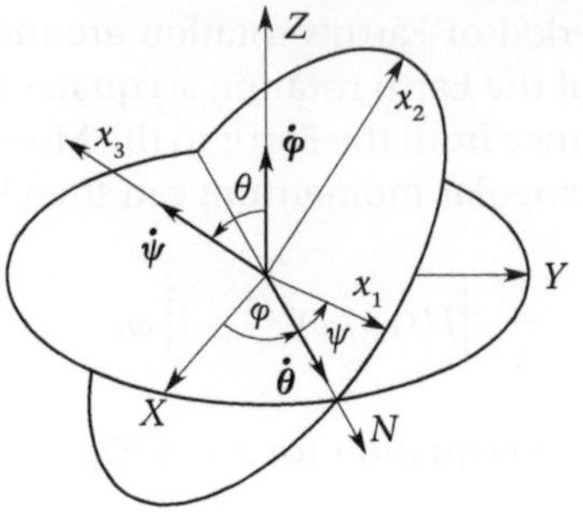

Figure 88 Euler angles

To express in terms of these angles and their time derivatives the kinetic energy of rotational of a rigid body, let us write components of the angular velocity

$$\boldsymbol{\Omega} = \dot{\boldsymbol{\varphi}} + \dot{\boldsymbol{\theta}} + \dot{\boldsymbol{\psi}}$$

in the frame K, in which the inertia tensor does not depend on time. It is easy to do this, taking into account that the polar and azimuth angles in this system for the vector $\dot{\boldsymbol{\varphi}}$ are θ and $(\pi/2) - \psi$, for vector $\dot{\boldsymbol{\theta}}$ are equal to $\pi/2$ and $(-\psi)$, and that the vector $\dot{\boldsymbol{\psi}}$ is directed along the x_3 axis (see Fig. 88). As a result, we get

$$\begin{aligned} \Omega_1 &= \dot{\varphi} \sin\theta \sin\psi + \dot{\theta} \cos\psi, \\ \Omega_2 &= \dot{\varphi} \sin\theta \cos\psi - \dot{\theta} \sin\psi, \\ \Omega_3 &= \dot{\varphi} \cos\theta + \dot{\psi}. \end{aligned} \tag{56.1}$$

If the axes of the system K be directed along the principal axes of inertia, then the kinetic energy (53.13) of the rotation of a rigid body,

$$T = \frac{1}{2}\left(I_1\Omega_1^2 + I_2\Omega_2^2 + I_3\Omega_3^2\right), \tag{56.2}$$

taking into account (1), is also expressed in terms of the Euler angles. In particular, the kinetic energy of a symmetrical top (we choose the axes $x_1x_2x_3$ so that $I_1 = I_2$) reads

$$T = \frac{1}{2}\left[I_1(\dot{\theta}^2 + \dot{\varphi}^2 \sin^2\theta) + I_3(\dot{\varphi}\cos\theta + \dot{\psi})^2\right]. \tag{56.3}$$

Expressions (2) and (3) can now be used to write the Lagrangian L. For free motion $L = T$; in this case the generalized momentum

$$p_\psi = \frac{\partial L}{\partial \dot{\psi}} = I_3\Omega_3 \tag{56.4}$$

is equal to the projection M_3 of the angular momentum on the axis x_3, the generalized momentum

$$p_\varphi = \frac{\partial L}{\partial \dot{\varphi}} = I_1\Omega_1 \sin\theta \sin\psi + I_2\Omega_2 \sin\theta \cos\psi + I_3\Omega_3 \cos\theta \tag{56.5}$$

is equal to the projection M_Z of the angular momentum on the OZ axis, and the generalized momentum

$$p_\theta = \frac{\partial L}{\partial \dot{\theta}} = I_1\Omega_1 \cos\psi - I_2\Omega_2 \sin\psi \tag{56.6}$$

is equal to the projection of the angular momentum on the line of nodes. The Lagrangian function for the free motion of an asymmetrical top does not depend on the angle φ, so $p_\varphi = M_Z$ is conserved. For free motion of a symmetrical top, there is additionally conserved momentum $p_\psi = M_3$, since in this case the Lagrangian function does not depend on the angle ψ.

Supplements

A Elements of the calculus of variations

The simplest example of variational problems is finding the shortest curve between two given points $A(x_1, y_1)$ and $B(x_2, y_2)$ on the xy-plane. This amounts to finding a function $y(x)$ that minimizes the integral

$$l = \int_A^B dl = \int_{x_1}^{x_2} \sqrt{1 + (dy/dx)^2}\, dx, \quad (dl)^2 = (dx)^2 + (dy)^2. \tag{A.1}$$

The function $y(x)$ must also satisfy the boundary conditions

$$y(x_1) = y_1, \;\; y(x_2) = y_2.$$

For this particular case, the solution is well-known to be a straight line given by

$$y(x) = y_1 + \frac{y_2 - y_1}{x_2 - x_1}(x - x_1). \tag{A.2}$$

In a more general form, variational problems are formulated as follows: let there be some class of functions $\tilde{y}(x)$ such that all of them pass through the points $A(x_1, y_1)$ and $B(x_2, y_2)$, i.e. they satisfy the boundary conditions $\tilde{y}(x_1) = y_1$ and $\tilde{y}(x_2) = y_2$. Among these functions, we seek to find a function $y(x)$ that makes the integral

$$J = \int_{x_1}^{x_2} f(y, y', x)dx, \quad y' = \frac{dy(x)}{dx} \tag{A.3}$$

take an extreme value.

Let us derive an equation that such a function $y(x)$ must satisfy. The curves, which we consider as taking part in the 'competition', are functions of the form

$$\tilde{y}(x) = y(x) + \varepsilon h(x), \quad h(x_1) = h(x_2) = 0, \tag{A.4}$$

where ε is a small parameter and $h(x)$ is sufficiently smooth (continuous together with its derivative) function (Fig. 89). The quantity $\varepsilon h(x)$ is called the *variation* of the function $y(x)$ and is denoted as

$$\delta y(x) \equiv \varepsilon h(x).$$

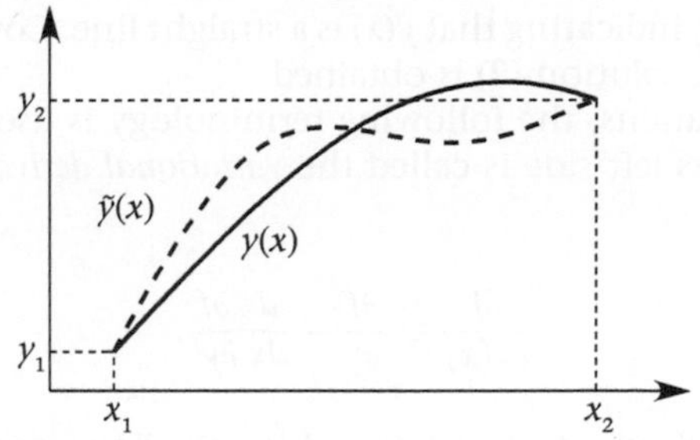

Figure 89 On the derivation of the Euler equation for the simplest variational problem

When substituting the function y with $\tilde{y}$ in Eq. (3), we get a quantity

$$J(\varepsilon) = \int_{x_1}^{x_2} f(y + \varepsilon h, y' + \varepsilon h', x)dx, \; h' = \frac{dh(x)}{dx},$$

which is a function of the parameter ε.

By virtue of assumption that $y(x)$ provides an extreme value of $J(\varepsilon)$, the point $\varepsilon = 0$ for $J(\varepsilon)$ must be an extreme point, so

$$\left.\frac{dJ(\varepsilon)}{d\varepsilon}\right|_{\varepsilon=0} = \int_{x_1}^{x_2} \left(\frac{\partial f}{\partial y} h + \frac{\partial f}{\partial y'} h' \right) dx = 0.$$

Integrating the second term under the integral by parts and taking into account that $h(x_1) = h(x_2) = 0$, we get

$$\int_{x_1}^{x_2} \left(\frac{\partial f}{\partial y} - \frac{d}{dx}\frac{\partial f}{\partial y'} \right) h(x)dx + \left.\frac{\partial f}{\partial y'} h(x)\right|_{x_1}^{x_2} = \int_{x_1}^{x_2} \left(\frac{\partial f}{\partial y} - \frac{d}{dx}\frac{\partial f}{\partial y'} \right) h(x)\, dx = 0. \tag{A.5}$$

The following statement (*the main lemma calculus of variations*) is true. If $g(x)$ is continuous and

$$\int_{x_1}^{x_2} g(x)h(x)dx = 0$$

for any continuous function $h(x)$ with continuous derivative and such that $h(x_1) = h(x_2) = 0$, then $g(x) = 0$ on interval $x \in (x_1, x_2)$. (Prove it!)

Using this lemma, we obtain from (5) the necessary condition that $y(x)$ gives the extremal value of the integral (3):

$$\frac{\partial f}{\partial y} - \frac{d}{dx}\frac{\partial f}{\partial y'} = 0. \tag{A.6}$$

For example (1), the following equation is given in the form

$$\frac{d}{dx}\frac{y'}{\sqrt{1 + y'^2}} = 0,$$

which implies $y' = \text{const}$, indicating that $y(x)$ is a straight line. Considering the conditions $y(x_1) = y_1, y(x_2) = y_2$, the solution (2) is obtained.

In the calculus of variations, the following terminology is used. Equation (6) is called the *Euler equation*, and its left side is called the *variational derivative* of J with respect to $y(x)$ and is denoted as

$$\frac{\delta J}{\delta y(x)} \equiv \frac{\partial f}{\partial y} - \frac{d}{dx}\frac{\partial f}{\partial y'}. \tag{A.7}$$

Variation (or, more precisely, the first variation) J is called the quantity δJ defined by the relation

$$\delta J \equiv \int_{x_1}^{x_2} \frac{\delta J}{\delta y(x)}\,\delta y(x)\,dx\,. \tag{A.8}$$

Similarly, one can set up and solve the problem of determining the extremum of integral

$$J = \int_{x_1}^{x_2} f(y_1,\ldots,y_s;\ y_1',\ldots,y_s', x)\,dx, \tag{A.9}$$

depending on many unknown functions $y_i(x)$, which are independent of each other and have variations that satisfy the conditions $\delta y_i(x_1) = \delta y_i(x_2) = 0\,,\ i = 1,\ldots,s$. The necessary extremum condition (6) must be satisfied with respect to each of these functions:

$$\frac{\delta J}{\delta y_i(x)} \equiv \frac{\partial f}{\partial y_i} - \frac{d}{dx}\frac{\partial f}{\partial y_i'} = 0, \quad i = 1,2,...,s. \tag{A.10}$$

B Systems with constraints

B.1 *Systems with ideal holonomic constraints*

Let us now consider the motion of systems with constraints in greater detail. Recall the definitions given at the end of § 15. We assume that the bodies under study, whose motion is being analysed, consist of N 'material points' (or 'particles'), and the motion of these particles may be limited by rods, surfaces, and so on. If all these constraints are expressed by conditions of the form

$$F_\alpha(\mathbf{r}_1,...,\mathbf{r}_N,t) = 0, \quad \alpha = 1,...,n, \tag{B.1}$$

where F_α is a function of the particle coordinates and time only, the motion of this system is said to be limited by n *holonomic constraints*. For example, in the case of a pendulum of variable length considered in § 15, the conditions (1) are reduced to one equation

$$F_1(\mathbf{r},t) \equiv r - l(t) = 0.$$

We assume that the conditions (1) are satisfied because particles are affected by reaction forces $\mathbf{R}_a$ due to constraints, as well as other potential forces

$$\mathbf{F}_a = -\frac{\partial U}{\partial \mathbf{r}_a}.$$

These constraints are regarded as *ideal* if the total work of all reaction forces is equal to zero for any particle displacement $\delta\mathbf{r}_a$, provided that the constraints (1) are not violated:

$$\sum_{a=1}^{N} \mathbf{R}_a \cdot \delta\mathbf{r}_a = 0\,. \tag{B.2}$$

Note that we are not referring here to the displacements taking place during the actual motion of the system, but rather to the displacements that do not violate the conditions (1) taken for a fixed value of time t.

In Newton's mechanics, the motion of particles is determined by equations

$$m_a\ddot{\mathbf{r}}_a = -\frac{\partial U}{\partial \mathbf{r}_a} + \mathbf{R}_a\,,\quad a = 1, \ldots, N, \tag{B.3}$$

which together with conditions (1) allow us to find both the law of motion $\mathbf{r}_a(t)$ and the forces $\mathbf{R}_a(t)$.

It should be noted that our system has $s = 3N - n$ degrees of freedom, taking into account constraint conditions (1), and exactly as many generalized coordinates as needed to describe its motion completely. Although the equations of motion for these generalized coordinates can be derived from equations (2), (3), and conditions (1), this approach is laborious. Therefore, we will utilize the advantages of the Lagrangian approach to derive these equations.

Conditions (1) for a system of N material points separate a subspace K_0 of $s = 3N - n$ dimensions in space K, where each point representing the system configuration can only move. Reaction forces depicted in space K are orthogonal to subspace K_0, complying with the ideal constants condition stated above. Introducing n generalized coordinates:

$$\tilde{q}_\alpha = F_\alpha(\mathbf{r}_1, \ldots, \mathbf{r}_N, t), \quad \alpha = 1, \ldots, n,$$

the remaining generalized coordinates are denoted by q_i, $i = 1, \ldots, s$. The subspace K_0 is defined by n conditions

$$\tilde{q}_\alpha = 0, \quad \alpha = 1, \ldots, n; \tag{B.4}$$

in other words, q_i are coordinates in K_0. Just like in the pendulum example in § 15, we can introduce an auxiliary system that has 'disabled' constraints but an additional extremely 'rigid' potential energy:

$$\tilde{U}(\tilde{q}_1, \ldots, \tilde{q}_n) = \sum_\alpha \tilde{U}_\alpha(\tilde{q}_\alpha)\,, \tag{B.5}$$

where $\tilde{U}_\alpha(\tilde{q}_\alpha)$ is a function of the same form, as in the pendulum example. The resulting forces $(-\partial\tilde{U}/\partial\tilde{q}_\alpha)$, as well as the reaction forces, are orthogonal to the subspace K_0.

As a matter of fact, we accept that, during the motion of a mechanical system, there arise the forces which just ensure the fulfilment of the constraint conditions. Although these forces are caused by the deformations of the bodies, those deformations are quite small, so the associated displacements and velocities can be disregarded.

The Lagrange function of the auxiliary system has the form

$$\tilde{L} = L_1(q, \tilde{q}, \dot{q}, \dot{\tilde{q}}, t) - \tilde{U}(\tilde{q}), \tag{B.6}$$

where L_1 denotes the Lagrangian function without considering constraints. The Lagrangian equations for coordinates $\tilde{q}_\alpha$ contain the quantities $R_\alpha = -\partial\tilde{U}/\partial\tilde{q}_\alpha$ that play the role of generalized forces of the constraint reaction. The transition to the original system with constraints involves declaring the coordinates $\tilde{q}_\alpha$, according to (4), as given and the quantities R_α as unknown. While writing the Lagrangian equations for coordinates q_i, it is possible to discard the term $\tilde{U}$ in (6) and substitute $\tilde{q}_\alpha = 0$, $\dot{\tilde{q}}_\alpha = 0$, which entails using the Lagrangian:

$$L(q, \dot{q}, t) = L_1(q, 0, \dot{q}, 0, t) = T(q, \dot{q}, t) - U(q, t)\,, \tag{B.7}$$

where $T(q, \dot{q}, t)$ denotes kinetic energy and $U(q, t)$ denotes potential energy.

In essence, for a system with ideal holonomic constraints, we can readily select the generalized coordinates, taking the constraints into account, and express the Lagrangian function in terms of them. In this manner, the desired equations of motion take the following form:

$$\frac{d}{dt}\frac{\partial L(q, \dot{q}, t)}{\partial \dot{q}_i} = \frac{\partial L(q, \dot{q}, t)}{\partial q_i}, \quad i = 1, 2, ..., s = 3N - n.$$

We leave it to the reader to verify that the same equations could also be derived from Eqs. (2) and (3), taking into account conditions (1). It can be done, for example, as follows. We express the reaction force of constraints from equation (3),

$$\mathbf{R}_a = m_a \ddot{\mathbf{r}}_a + \frac{\partial U}{\partial \mathbf{r}_a},$$

and substitute this expression into equation (2):

$$\sum_{a=1}^{N} \left[m\ddot{\mathbf{r}}_a + \frac{\partial U}{\partial \mathbf{r}_a} \right] \cdot \delta \mathbf{r}_a = 0\,.$$

Next, the vectors $\mathbf{r}_a$ and displacements $\delta\mathbf{r}_a$ can be expressed in terms of generalized coordinates q_i and the corresponding displacements δq_i. By doing so, we can use the independence of δq_i variations.

Among the systems with ideal holonomic constraints is the absolutely rigid body. This refers to a set of particles where the distances between them remain unchanged. Experience shows that using this model to describe the motion of many bodies is applicable. The position of a rigid body can be determined with just six coordinates (see § 52).

When rigid bodies are in contact, the constraints are ideal in two limiting cases: if friction can be neglected, or if slipping is impossible. In both cases, the work of friction forces is zero. Conditions that restrict the possible motions of bodies may also contain velocities. For example, consider a cylinder rolling without slipping along a fixed plane (Fig. 90). In this case, the condition is equal to the zero velocity of the point touching the plane:

$$\dot{x} + a\dot{\varphi} = 0,$$

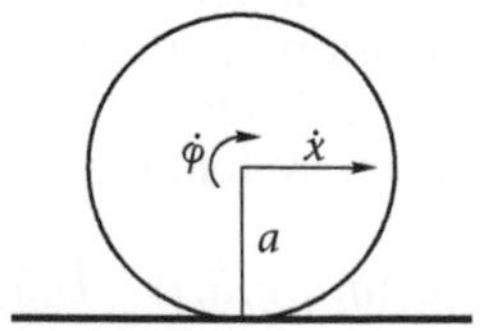

Figure 90 A cylinder rolls without slipping on a stationary plane

where x is the axis coordinate and φ is the rotation angle of the cylinder. This condition can be integrated:

$$x + a\varphi = \text{const},$$

and the constraints turns out to be holonomic.

For a ball or disk rolling on a plane, the constraint condition

$$\mathbf{V} + [\mathbf{\Omega}, \mathbf{a}] = 0 \tag{B.8}$$

cannot be integrated (here **a** is the radius directed from the centre of the disk to the point of contact). Such a constraint is called *non-holonomic*. The outlined scheme for discovering a system's motion is not applicable in this case. The description of how to find the equations of motion for systems with non-holonomic constraints will be considered in section B.3.

B.2 *Reaction forces of constraints*

After solving the problem of the system motion, one can, if necessary, determine the reaction forces. This problem requires additional analysis, and while there are many different possibilities, we will confine ourselves to this one example. By using changes in lengths of rods as coordinates $\tilde{q}$, we can apply the same method as before to find the forces that stretch the rods during motion of the system.

We will consider a motion in the plane of the system of four small bodies (which we will call 'particles') A, B, C, and D, connected by five rigid rods (Fig. 91), hinged to the bodies (so that the angles between the rods can freely change if any of them are deleted). The system has three degrees of freedom (say, two coordinates of one of the particles and the angle that determines the direction of one of the sides). The constraints of the system express the constancy of the lengths of the rods. However, these restrictions can also be expressed using the condition $BD = \text{const}$ instead of $AC = \text{const}$. By using this new condition, we obtain the reaction force supposedly directed along BD, while losing the reaction force of the existing rod AC. In this case, the reaction forces of the remaining rods turn out to be not true, but appear as if the rod AC was really removed and the BD rod was introduced into the system.

It is important to note that all of these different possibilities for describing the constraints and their actual creation are completely unimportant for solving the problem of system motion using Lagrangian equations. However, by correctly choosing equations that express ideal holonomic constraints, it is possible to obtain the reaction forces of the constraints using the method of indefinite Lagrangian multipliers to determine conditions for the extremum of the action.

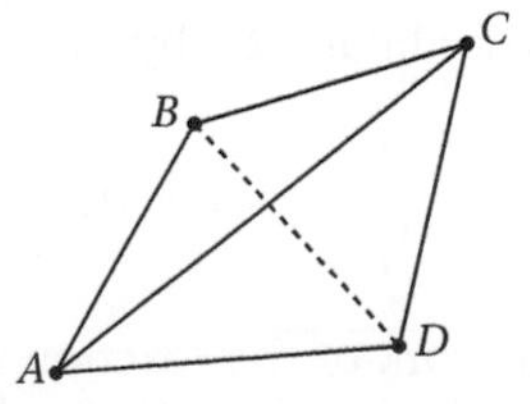

Figure 91 Bodies with constraints

B.3 *Indefinite Lagrangian multipliers. Ideal non-holonomic constraints*

Let us return to the system with ideal holonomic constraints. We will assume that we did not use all of the equations of connection and eliminated fewer coordinates than we could.

Let the Lagrangian function depend on $s+n$ coordinates and as many velocities, $L = L(q_1, \ldots, \dot{q}_{s+n}, t)$, and let there be n ideal holonomic constraints:

$$F_\alpha(q_1, \ldots, q_{s+n}, t) = 0, \quad \alpha = 1, ..., n. \tag{B.9}$$

Differentiating (9) with respect to time, we can represent the conditions of the constraints as

$$\sum_{i=1}^{s+n} c_{\alpha i}\, \dot{q}_i + c_\alpha = 0, \tag{B.10}$$

where

$$c_{\alpha i} = \frac{\partial F_\alpha}{\partial q_i}, \quad c_\alpha = \frac{\partial F_\alpha}{\partial t}. \tag{B.11}$$

In this case, the condition for the extremum action that defines the equations of motion for a system of bodies can be obtained using indefinite Lagrangian multipliers, rather than excluding n extra variables.

The method for determining the extremum in a variational problem using indefinite Lagrangian multipliers is similar to that for determining the extremum of a function of several variables in the presence of connections between them. An auxiliary function is introduced that includes unknown functions of time $\lambda_\alpha(t)$,

$$\tilde{L}(q_1, \ldots, \dot{q}_{s+n}, t) = L(q_1, \ldots, \dot{q}_{s+n}, t) + \sum_{\alpha=1}^{n} \lambda_\alpha(t)\, F_\alpha(q_1, \ldots, q_{s+n}, t), \tag{B.12}$$

write 'action' $\tilde{S} = \int \tilde{L} dt$ for it and get Lagrangian equations, considering all $s+n$ coordinates as independent. Thus, we allow functions other than those we need or which satisfy the condition (9) to participate in the 'competition'. These $s+n$ equations should be supplemented with n constraint equations (9), thereby excluding temporarily added dependencies $q_i(t)$. As a result, we have just $s+2n$ equations for all coordinates and Lagrangian multipliers:

$$\frac{d}{dt}\frac{\partial L}{\partial \dot{q}_i} - \frac{\partial L}{\partial q_i} = \sum_{\alpha=1}^{l} \lambda_\alpha c_{\alpha i}. \tag{B.13}$$

The sums on the right-hand sides of the equations can be considered as generalized forces acting 'in directions' of the corresponding generalized coordinates. An additional condition can be either equalities (9) or (10). Note that conditions (10) (for a fixed time) limit the possible variations of coordinates δq_i by the relations:

$$\sum_{i=1}^{s+n} c_{\alpha i}\, \delta q_i = 0. \tag{B.14}$$

These conditions mean that the work of the reaction forces of the constraints becomes zero if the variations of the coordinates δq_i satisfy relations (14), and this ensures the ideality of the constraints.

The fact that the left side of equality (10) is the complete derivative with respect to time is equivalent to the relations

$$\frac{\partial c_{\alpha i}}{\partial q_j} = \frac{\partial c_{\alpha j}}{\partial q_i}, \quad \frac{\partial c_{\alpha i}}{\partial t} = \frac{\partial c_{\alpha}}{\partial q_i}, \tag{B.15}$$

for all α, i, j.

Linear ideal non-holonomic constraints can also be expressed by relations like (10) or (14). Conditions for the ideality of the non-holonomic constraints are expressed in terms of coordinate variations in the same way as for holonomic ones. These relations, for both holonomic and non-holonomic constraints, mean that under the 'unauthorized' displacement of bodies, such forces (perhaps, very large ones) would appear that could make such displacement impossible. On the other hand, the additional forces do not occur under 'authorized' displacements. Moreover, the equalities of the form (15) for the corresponding coefficients are not fulfilled in the case of non-holonomic constraints; therefore, the conditions for constraints cannot be reduced to the form (9).

Note, however, that relations (15) are connected with the smallness of the second order in variations. These equalities mean that the work δA_{12} on the way

$$(q_1, q_2) \to (q_1 + \delta q_1, q_2) \to (q_1 + \delta q_1, q_2 + \delta q_2)$$

for the α-th constraint and work δA_{21} on the way

$$(q_1, q_2) \to (q_1, q_2 + \delta q_2) \to (q_1 + \delta q_1, q_2 + \delta q_2)$$

is the same up to $\delta q_1 \delta q_2$ inclusive.

Indeed,

$$\delta A_{12} = \lambda_\alpha [c_{\alpha 1}(q_1, q_2, \ldots)\delta q_1 + c_{\alpha 2}(q_1 + \delta q_1, q_2, \ldots)\delta q_2] =$$

$$= \lambda_\alpha [c_{\alpha 1}(q_1, q_2, \ldots)\delta q_1 + c_{\alpha 2}(q_1, q_2, \ldots)\delta q_2 + \frac{\partial c_{\alpha 2}(q_1, q_2, \ldots)}{\partial q_1}\delta q_1 \delta q_2],$$

while δA_{21} is obtained by replacing $1 \rightleftarrows 2$, so that

$$\delta A_{12} - \delta A_{21} = \lambda_\alpha \left(\frac{\partial c_{\alpha 2}}{\partial q_1} - \frac{\partial c_{\alpha 1}}{\partial q_2} \right).$$

For the holonomic constraint according to (15), we have $\delta A_{12} - \delta A_{21} = 0$ up to the second order inclusive. This exact condition is the one that distinguishes a holonomic constraint from a non-holonomic one by providing the possibility to represent the reaction forces via the potential energy.

When defining the equations of motion and force, we only operated with expressions linear in δq. Therefore, equations (13) with additional conditions of the form (10) are also valid for the mechanical systems with non-holonomic constraints.

For a system with n holonomic constraints, one can exclude n generalized coordinates from the Lagrangian function. For a system with non-holonomic connections, such a decrease in the number of generalized coordinates is essentially impossible.

As an example of a non-holonomic constraint, consider the constraint condition for a wheel that can roll on a flat surface without slipping. It is clear that such a wheel can be rolled under the conditions of constraint and then returned to the starting point. However, the spoke of the wheel, which looked down in the initial state of the wheel, will be directed somehow differently at the end of motion.

C Hill equation, Mathieu equation, and parametric resonance

Basic information about parametric resonance can be found in § 36. This Supplement provides a more detailed and consistent theory of this interesting phenomenon.

C.1 *General properties of the Hill equation*

The basic equation for study of a parametric resonance has the form (36.1)

$$\ddot{x} + \omega^2(t)\, x = 0\,, \tag{C.1}$$

where the frequency $\omega(t)$ changes periodically in time

$$\omega(t+T) = \omega(t). \tag{C.2}$$

Equation (1) with periodic dependence of frequency on time (2) is called the *Hill equation*. When constructing approximate solutions, it is useful to know as many properties of the exact solutions as possible. Here is a list of some simple and useful properties of the solutions of the Hill equations. To do this, the *Wronsky determinant* is introduced. Let $x_1(t)$ and $x_2(t)$ be two solutions of Eq. (1). By definition, the Wronsky determinant is equal to:

$$W(x_1, x_2) = \begin{vmatrix} x_1 & x_2 \\ \dot{x}_1 & \dot{x}_2 \end{vmatrix} = x_1(t)\, \dot{x}_2(t) - x_2(t)\, \dot{x}_1(t).$$

1. By differentiating the Wronsky determinant with respect to time and getting rid of the second derivatives by using equation (1), we easily find that

$$\frac{dW}{dt} = 0\,,$$

i.e. W does not depend on time.

2. If $x(t)$ is a solution of Eq. (1), then so is $x(t+T)$. This property is proved by substituting into the equation and using the periodicity of ω. From here, the following can be concluded: if $x_1(t)$ and $x_2(t)$ are linearly independent and form a basis in the space of solutions, then the following equalities hold:

$$x_1(t+T) = a_{11}x_1(t) + a_{12}x_2(t),$$
$$x_2(t+T) = a_{21}x_1(t) + a_{22}x_2(t).$$

The matrix of constant coefficients, denoted by a_{ij}, is defined by both Eq. (1) and the choice of basic solutions. A basis in the space of solutions can be chosen such that the matrix becomes diagonal. In other words, there exist solutions $x_1(t)$ and $x_2(t)$ which, when shifted by the period T, are transformed by the following law:

$$x_1(t+T) = \mu_1\, x_1(t)\,, \quad x_2(t+T) = \mu_2\, x_2(t)\,, \tag{C.3}$$

where $\mu_{1,2}$ are the eigenvalues of the matrix a_{ij}.

3. The relation

$$W(x_1(t+T),\, x_2(t+T)) = \mu_1\,\mu_2\, W(x_1(t),\, x_2(t))$$

can be proven valid for the given solutions (3). From this relation and property 1, we obtain:

$$\mu_1\,\mu_2 = 1. \tag{C.4}$$

4. The determinant of the matrix a_{ij} is equal to 1 due to Property 3. The eigenvalues of the matrix are given by:

$$\mu_{\pm} = d \pm \sqrt{d^2 - 1}\,, \quad d = \frac{1}{2}\,(a_{11} + a_{22})\,. \tag{C.5}$$

If $d^2 > 1$, then μ_+ and μ_- are real. One of them has a modulus greater than one, denoted as μ_1. It is important to note that this can be either $\mu_1 > 1$ (if $d>1$, in which case $\mu_1 = \mu_+$) or $\mu_1 < -1$ (if $d < -1$, in which case $\mu_1 = \mu_-$). Therefore, the solution $x_1(t)$ given by Eq. (3) increases in absolute value over the period T. This phenomenon is referred to as *parametric resonance*. On the other hand, the second solution $x_2(t)$ decreases in modulus over the same period.

If $d^2 < 1$, the eigenvalues μ_1 and μ_2 are complex conjugates ($\mu_2 = \mu_1^*$) and lie on the circle of radius one ($|\mu_1| = |\mu_2| = 1$) in the complex plane. Solutions $x_{1,2}(t)$ turn out to be complex functions and can be chosen to be complex conjugate, i.e. $x_2(t) = x_1^*(t)$. The real solutions are then given by their linear combinations: $x_1(t) + x_2(t)$ and $[x_1(t) - x_2(t)]/i$. In this case, the motion of the oscillator represents *beats*.

5. Let $x_i(t)$ represent the constructed complex-valued solutions of Eq. (3). Every shift by T leads to multiplication of the solution by μ_i, thus changing the solution exponentially on average. It is obvious that $x_i(t)/\mu_i^{t/T}$ is a periodic function. Therefore, there exist two linearly independent complex solutions of the Hill equation, which can be written as:

$$x_1(t) = \mu_1^{t/T}\Pi_1(t), \quad x_2(t) = \mu_2^{t/T}\Pi_2(t)\,, \tag{C.6}$$

where $\Pi_1(t)$ and $\Pi_2(t)$ are purely periodic functions, and μ_1 and μ_2 are related by condition (4).

6. It is easy to generalize the above consideration and seek a solution for the equation that takes into account low linear friction:

$$\ddot{x} + 2\lambda\dot{x} + \omega^2(t)\,x = 0\,. \tag{C.7}$$

By substituting

$$x = e^{-\lambda t}\,\tilde{x}(t)\,, \tag{C.8}$$

we obtain the Hill equation (1) for the function $\tilde{x}(t)$.

C.2 *Mathieu equation*

Let us find out the conditions for the parametric resonance in the case when

$$\omega^2(t) = \omega_0^2(1 + h\cos\gamma t)\,, \tag{C.9}$$

where $\gamma = 2\pi/T$ and the constant $h \ll 1$. The equation

$$\ddot{x} + \omega_0^2(1 + h\cos\gamma t)\,x = 0 \tag{C.10}$$

(for arbitrary h and γ) is called the *Mathieu equation*.

When friction is taken into account, Eq. (10) is transformed into the equation of the form (7):

$$\ddot{x} + 2\lambda\dot{x} + \omega_0^2(1 + h\cos\gamma t)\,x = 0\,, \tag{C.10a}$$

which reduces to the Mathieu equation using the substitution (8).

For the Mathieu equation, the quantity μ_1 is a function of the parameters γ and h. The boundary, which separates the regions of stability and instability, is given by equations $\mu_1(\gamma, h) = -1$ and $\mu_1(\gamma, h) = 1$. These equations define curves in the (γ, h) plane, which we will call *neutral curves*. Let us construct the neutral curves for small h starting from the degenerate case $h = 0$. For $h = 0$, the frequency $\omega = \omega_0$ does not change at all and the solution is well-known:

$$x_1(t) = a\,e^{i\omega_0 t}\,.$$

On the other hand, nothing prevents us from assuming that the frequency ω changes with an arbitrary period $T = 2\pi/\gamma$, although with zero amplitude. In this case, we can represent this solution in the form:

$$x_1(t) = a\,e^{i\omega_0 t} = a\,e^{i(\omega_0-\gamma)t}\,e^{i\gamma t} = e^{i(\omega_0-\gamma)t}\,\Pi_1(t)\,,$$

where

$$\mu_1 = \exp\left[i(\omega_0 - \gamma)\frac{2\pi}{\gamma}\right]\,.$$

Assuming both ω_0 and γ to be positive, it becomes evident that $\mu_1 = \pm 1$ under condition

$$\frac{\omega_0}{\gamma} = \frac{n}{2}\,,$$

where $n \geq 1$ is an integer, while $\mu_1 = -1$ if n is odd and $\mu_1 = 1$ if n is even. For $h \ll 1$, it is natural to assume that the frequency γ lies close to those found discrete values on the neutral curve.

C.3 *Parametric resonance on the fundamental harmonic* $\gamma = 2\omega_0$

Let us consider the main parametric resonance with $n = 1$ and, respectively, at

$$\gamma = 2\omega_0 + \epsilon\,,$$

where the detuning ε is assumed to be small. It is convenient to present μ_1 (which in this case is close to (-1)) in the form

$$\mu_1 = -e^{sT} = e^{sT - i\pi}\,,$$

where s is an unknown real parameter equal to zero on the neutral curve and which is small and positive in the region of instability.

Looking for $\Pi(t)$ in the form of a Fourier series

$$\Pi(t) = \sum_{n=-\infty}^{+\infty} A_n\, e^{in\gamma t}$$

with unknown coefficients A_n, we conclude that the solution of equation (10) should be sought in the form

$$x(t) = e^{st} \sum_{n=-\infty}^{+\infty} A_n\, e^{i\,(2n-1)(\omega_0+\epsilon/2)\,t}\,. \tag{C.11}$$

Substituting series (11) into Eq. (10) and representing $\cos\gamma t$ as a half-sum of the exponents, we then equate to zero the coefficients for all harmonics. As a result, we obtain an infinite system of homogeneous linear algebraic equations for the coefficients A_n:

$$\left\{\omega_0^2 + [s + i(2n-1)(\omega_0 + \epsilon/2)]^2\right\} A_n = -\frac{1}{2}\, h\omega_0^2\,(A_{n-1} + A_{n+1})\,. \tag{C.12}$$

The increment s is determined from the condition that the determinant of this infinite system $(-\infty < n < +\infty)$ is equal to zero.

We will look for a solution using the smallness of the parameter h. In the zero approximation (for $h \to 0$), series (11) is reducible to the usual free oscillations with a constant frequency ω_0:

$$x(t) \to a\,\cos(\omega_0 t + \varphi) = A_+\, e^{i\omega_0 t} + A_-\, e^{-i\omega_0 t}\,,\quad A_\pm = \frac{1}{2}\, a\, e^{\pm i\varphi}\,. \tag{C.13}$$

Comparing Eqs. (11) and (13), we find that $s = \epsilon = 0$ in the zero approximation and that all amplitudes, except A_0 and A_1, are equal to zero. Substituting the zero approximation into the right side of Eq. (12), we find further that in the first approximation in h only coefficients $A_{-1} = \frac{1}{16}\, hA_0$ and $A_2 = \frac{1}{16}\, hA_1$ are not equal to zero.

As a result, leaving only the terms of the first order in Eq. (12), we get a system of two equations:

$$\begin{aligned} 2\,(\epsilon + 2is)\,A_0 - h\omega_0\,A_1 &= 0,\\ 2\,(\epsilon - 2is)\,A_1 - h\omega_0\,A_0 &= 0. \end{aligned} \tag{C.14}$$

Equating the determinant of this system to zero, we obtain the following expression for the increment:[1]

$$s = \frac{1}{14}\sqrt{(h\omega_0)^2 - 4\epsilon^2}.$$

Hence, recalling the definition of ε, we find the neutral curve $(s = 0)$:

$$\gamma = 2\omega_0 \pm \frac{1}{2}\, h\omega_0\,.$$

The region of instability starts from $h = 0$ and lies between diverging dotted lines in Fig. 92; the width of this region in frequency grows linearly with h.

[1] The second solution corresponds to another sign s, which is why the motion of the considered oscillator that satisfies the arbitrary initial conditions has the form of Eq. (36.6).

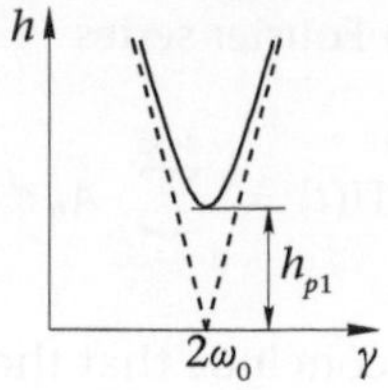

Figure 92 Parametric instability region around $2\omega_0$

The solution of the equation with friction (10*a*) differs from the one found by an additional factor $\exp(-\lambda t)$ (see Eq. (8)), so for the increment in this case, we have

$$s = -\lambda + \frac{1}{14}\sqrt{(h\omega_0)^2 - 4\epsilon^2}\,.$$

The excitation of oscillations is possible only when

$$h > h_{p1} = \frac{4\lambda}{\omega_0}\,;$$

in this case, the region of parametric instability is located above the solid curve in Fig. 92.

C.4 *Parametric resonance at* $\gamma = \omega_0$

Let us find the conditions for resonance at $n=2$ and $\gamma = \omega_0 + \epsilon$. We present μ_1, which is close to 1, as $\mu_1 = e^{sT}$. We are looking for the solution of Eq. (10) in the form

$$x(t) = e^{st} \sum_{n=-\infty}^{+\infty} A_n\, e^{i\,n(\omega_0+\epsilon)\,t} \tag{C.15}$$

and obtain a system of equations for the coefficients A_n

$$\left\{\omega_0^2 + [s + in(\omega_0+\epsilon)]^2\right\} A_n = -\frac{1}{2}\,h\omega_0^2\,(A_{n-1} + A_{n+1})\,, \tag{C.16}$$

which differs from system (12) only by the replacements $(2n-1) \to n$ and $\epsilon/2 \to \epsilon$.

Next, we repeat the actions described in the previous section. Specifically, by comparing Eqs. (15) and (13), it becomes evident that $s = \epsilon = 0$ in the zero in h approximation and that all amplitudes, except A_{-1} and A_1, are equal to zero. Substituting the zero approximation into the right side of Eq. (16), we find that in the first in h approximation, only these three coefficients

$$A_{\pm 2} = \frac{1}{16}A_{\pm 1}\,,\ \ A_0 = \frac{1}{2}\,h\,(A_{-1} + A_1) \tag{C.17}$$

do not equal zero. After that, leaving only the terms of the first order in Eq. (16), we obtain a system of two trivial equations:

$$(\epsilon + is)\,A_{-1} = 0\,,\ \ (\epsilon - is)\,A_1 = 0\,,$$

from which it follows that $s = \epsilon = 0$ even with the account the terms of the first order in h inclusive.

Therefore, it is necessary to take into account the second order in h. This leads to the following system of equations:

$$\begin{aligned} 4(\epsilon + is) A_{-1} &= h\omega_0 (A_{-2} + A_0) , \\ 4(\epsilon - is) A_1 &= h\omega_0 (A_2 + A_0) . \end{aligned} \tag{C.18}$$

Then we express A_{-2}, A_0, A_2 in terms of A_{-1} and A_1 using Eq. (17) and transform Eq. (18) in such a way that they contain only A_{-1} and A_1. By equating the determinant of the obtained system of two equations to zero, we get the expression for the increment

$$s = \frac{1}{8} \sqrt{h^4 \omega_0^2 - \left(8\epsilon + \frac{2}{3} h^2 \omega_0\right)^2} .$$

The neutral curve turns out to be asymmetric, and its left and right branches are expressed as follows

$$\gamma_l = \omega_0 - \frac{5}{24} h^2 \omega_0 ,$$

$$\gamma_n = \omega_0 + \frac{1}{24} h^2 \omega_0 .$$

The region of instability starts from $h = 0$ and is located between the branches depicted by dotted lines on Fig. 93; the width of this region increases with frequency proportional to h^2.

If there is friction, then the increment is equal to

$$s = -\lambda + \frac{1}{8} \sqrt{h^4 \omega_0^2 - \left(8\epsilon + \frac{2}{3} h^2 \omega_0\right)^2} ,$$

the minimum on the neutral curve $h = h(\gamma)$ is shifted to the point

$$\gamma_{min} = \omega_0 - \frac{1}{12} h^2 \omega_0,$$

and the threshold value h in it is equal to

$$h_{p2} = \sqrt{\frac{8\lambda}{\omega_0}} .$$

Thus, the excitation of oscillations is possible only when $h > h_{p2}$ and the region of parametric instability is located above the solid curve in Fig. 93.

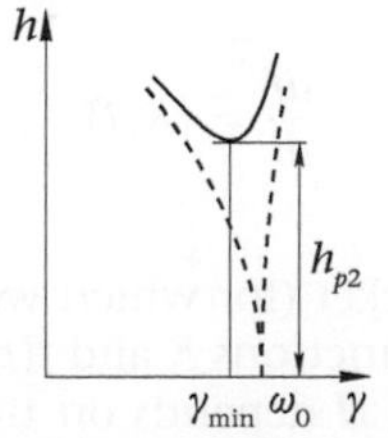

Figure 93 Parametric instability region around ω_0

Since the friction was assumed to be small ($\lambda \ll \omega_0$), it follows that $h_{p2}/h_{p1} \gg 1$. In this case, it becomes difficult to get even the second resonance. As for the thresholds for higher resonances, they will be even larger, and with increasingly smaller widths.

D Generalization of canonical transformations

D.1 *Time and energy as canonical variables*

Here a quick reminder is due. The Lagrangian equations admit transformations of both coordinates and time. In this section we will show that it is possible to generalize the canonical transformations so that they also allow transformation of the time.

In what follows, the canonical variables $q_i, p_i, ; i = 1, \ldots, s$ are denoted in shorthand as q, p. The Hamiltonian H is also assumed to be time-dependent $H = H(q, p, t)$. Besides the usual canonical variables, a couple more variables q_0, p_0 are introduced along with the new Hamiltonian

$$K(q, p, q_0, p_0) = H(q, p, q_0) + p_0$$

(which is obviously independent of time). Let us consider the Hamiltonian equations obtained with the help of the Hamiltonian K. First of all, note that

$$\dot{q}_0 = \frac{\partial K}{\partial p_0} = 1,$$

whence it follows that $q_0 = t + \text{const}$. Therefore, the variable q_0 can be identified with the time (putting without loss of generality a constant equal to zero). Similarly, the equation

$$\dot{p}_0 = -\frac{\partial K}{\partial q_0} = -\frac{\partial H}{\partial t} = -\frac{dE}{dt}$$

allows us to identify p_0 with $(-E)$. The remaining Hamiltonian equations

$$\dot{p}_i = -\frac{\partial K}{\partial q_i} = -\frac{\partial H}{\partial q_i}, \quad \dot{q}_i = \frac{\partial K}{\partial p_i} = \frac{\partial H}{\partial p_i}, \quad i = 1, \ldots, s$$

actually do not change.

It is easy to check that for an arbitrary function $f(q, p, t)$ we have

$$\frac{df}{dt} = \{K, f\}, \tag{D.1}$$

where the generalized Poisson bracket (for which we keep the same notation) includes terms obtained by differentiating functions K and $f(p, q, q_0)$ with respect to p_0 and q_0.

Even if the original Hamiltonian H depends on time, the Hamiltonian K satisfies the conservation law $K = 0$ (which reduces to $H(q, p, t) = E(t)$).

D.2 *Canonical transformations involving time and energy*

Now the generalization of canonical transformations can be easily done (compare with Eq. (43.2)):

$$\sum_{i=0}^{s} p_i dq_i - \sum_{i=0}^{s} P_i dQ_i = dF \tag{D.2}$$

(terms with $i = 0$ have been included in the sums). Since time is explicitly excluded in the Hamiltonian K, the appropriate substitution needs to be made in K when transitioning to new variables.

The proof of the invariance of the Poisson brackets with respect to canonical transformations, given in section E.4 below, can also be applied to this case. Additionally, taking into account Eq. (1), the Hamiltonian equations conserve their form when written using the Hamiltonian $K(q(Q,P), p(Q,P))$.

Example. *Let us consider the Lorentz transformation for a free particles corresponding to the transition to a frame of reference that moves with velocity* V *in the direction of the* x*-axis:*

$$x = \gamma(x' + \beta x'), \quad t = \gamma(t' + \beta x'/c);$$

where

$$\beta = V/c, \quad \gamma = 1/\sqrt{1 - V^2/c^2}.$$

In a similar way, as is known, the momentum of the particle and its energy are transformed:

$$p_x = \gamma(p_x' + \beta E'/c), \; E = \gamma(E' + \beta p_x' c).$$

Denoting

$$q_0 = ct, \quad q_1 = x, \quad p_0 = -E/c, \quad p_1 = p_x,$$
$$Q_0 = ct', \quad Q_1 = x', \quad P_0 = -E'/c, \quad P_1 = p_x',$$

we get

$$q_0 = \gamma(Q_0 + \beta Q_1), \quad q_1 = \gamma(Q_1 + \beta Q_0);$$
$$p_0 = \gamma(P_0 - \beta P_1), \quad p_1 = \gamma(P_1 - \beta P_0).$$

It is now easy to verify that

$$\{q_0, q_1\}_{P,Q} = 0, \; \{q_0, p_0\}_{P,Q} = 1$$

etc. The equalities $\{q_0, y\}_{P,Q} = 0, \; \{q_0, p_y\}_{P,Q} = 0$, *etc. should also be confirmed. Therefore, the transformation in question is canonical (generalized in the manner indicated above).*

E Differential forms and canonical transformations

The use of the mathematical apparatus of differential forms makes dealing with some questions related to canonical transformations easier and more concise. However, developing this apparatus requires certain efforts. The following summary of this issue is very brief and fragmented. A more detailed and consistent exposition of the theory of external differential forms in relation to the problems of mechanics can be found, for example, in the book [4].

E.1 *Differential forms*

The change of variable $x = x(\xi)$ in the integral

$$\int f(x)\, dx$$

is reduced, as is well known, to substitutions

$$f = f(x(\xi))\,, \quad dx = \frac{dx}{d\xi}\, d\xi\,.$$

The situation is different when the two variables must be replaced

$$x = x(\xi, \eta), \quad y = y(\xi, \eta)$$

in the double integral

$$I = \int f(x, y)\, dxdy\,.$$

In this case, the 'element of area'

$$dx\, dy \tag{E.1}$$

is replaced by the expression

$$D\, d\xi\, d\eta,$$

where

$$D = \frac{\partial(x, y)}{\partial(\xi, \eta)}$$

is the Jacobian of $xy \to \xi\eta$ transformation, but not the product of differentials

$$dx = \frac{\partial x}{\partial \xi} d\xi + \frac{\partial x}{\partial \eta} d\eta\,, \quad dy = \frac{\partial y}{\partial \xi} d\xi + \frac{\partial y}{\partial \eta} d\eta\,. \tag{E.2}$$

However, one can introduce a rule for the multiplication of differentials such that the transformation of the 'element of area' $dxdy$ is still reduced to substituting (2) into (1).

We introduce the so-called *external product of differentials*, denoted by

$$dx \wedge dy,$$

which can be manipulated like an ordinary product; however, when permuting the factors, it is necessary to change the sign:

$$dy \wedge dx = -dx \wedge dy \,. \tag{E.3}$$

It is also possible to multiply identical differentials. In this case, according to (3), zero will be obtained:

$$dx \wedge dx = 0.$$

Substitution (2) in (1)

$$\begin{aligned} dx \wedge dy &= \left(\frac{\partial x}{\partial \xi} d\xi + \frac{\partial x}{\partial \eta} d\eta\right) \wedge \left(\frac{\partial y}{\partial \xi} d\xi + \frac{\partial y}{\partial \eta} d\eta\right) = \\ &= \frac{\partial x}{\partial \xi}\frac{\partial y}{\partial \xi} d\xi \wedge d\xi + \frac{\partial x}{\partial \eta}\frac{\partial y}{\partial \xi} d\eta \wedge d\xi + \frac{\partial x}{\partial \xi}\frac{\partial y}{\partial \eta} d\xi \wedge d\eta + \frac{\partial x}{\partial \eta}\frac{\partial y}{\partial \eta} d\eta \wedge d\eta = \\ &= \left(\frac{\partial x}{\partial \xi}\frac{\partial y}{\partial \eta} - \frac{\partial x}{\partial \eta}\frac{\partial y}{\partial \xi}\right) d\xi \wedge d\eta = D d\xi \wedge d\eta \,, \end{aligned}$$

indeed, leads to the replacement of $dxdy$ by $Dd\xi d\eta$.

It is easy to generalize this rule to the case of any number of variables. The external product of n differentials

$$dx_1 \wedge dx_2 \wedge ... \wedge dx_n$$

is determined by the following condition: one can rearrange any neighbouring differentials, while changing the sign of the product.

When replacing

$$x_i = x_i(\xi_1, \xi_2, \dots, \xi_n), \quad i = 1, 2, \dots, n,$$

which corresponds to the replacement of differentials

$$dx_i = \sum_j \frac{\partial x_i}{\partial \xi_j} d\xi_j \,, \tag{E.4}$$

we obtain

$$dx_1 \wedge dx_2 \wedge \dots \wedge dx_n = \frac{\partial(x_1, x_2, \dots, x_n)}{\partial(\xi_1, \xi_2, \dots, \xi_n)} d\xi_1 \wedge d\xi_2 \wedge \dots \wedge d\xi_n \,. \tag{E.5}$$

Indeed, after substituting (4) into the left side of (5) and multiplying, we obtain the sum of products of n factors of the form $\partial x_i/\partial \xi_j$ with n different values of index i; each factor is the product of n differentials with all possible values of indices j:

$$d\xi_{j1} \wedge d\xi_{j2} \wedge \dots \wedge d\xi_{jn} \,.$$

However, the product of differentials is non-zero only if (i) all indices j in it are different, (ii) all these products of differentials are equal to each other or (iii) differ only in sign, depending on whether the transition from index sequence $j1, j2, \ldots, jn$ to the sequence $1, 2, \ldots, n$ has been done by an even number of permutations of neighbouring indices or an odd number. As a result, we actually obtain a description of the determinant composed of derivatives of $\partial x_i/\partial \xi_j$, that is, the Jacobian on the right side of equality (5).

The expression of the form

$$\sum_{i,j,\ldots,k} F_{i,j,\ldots,k}(x_1, x_2, \ldots, x_n)\,,\ dx_i \wedge dx_j \wedge \ldots dx_k,$$

in which each term contains the product of l differentials, is called the *external differential* l *-form* and is denoted ω^l. For example, the integrand of the integral over surfaces

$$\int \mathbf{A}\, d\mathbf{S}$$

can be viewed as a 2-form:

$$\omega^2 = A_x dy \wedge dz + A_y dz \wedge dx + A_z dx \wedge dy.$$

The *external differential of the* l *-form* is also introduced. By definition, this is the $(l+1)$-form

$$d\omega^l = \omega^{l+1} = \sum_{i,j,\ldots,k} dF_{i,j,\ldots,k} \wedge dx_i \wedge dx_j \wedge \ldots dx_k,$$

where $dF_{i,j,\ldots,k}$ is the usual differential of the function:

$$dF_{i,j,\ldots,k} = \sum_{m=1}^{n} \frac{\partial F_{i,j,\ldots,k}}{\partial x_m}\, dx_m\,. \tag{E.6}$$

When the external form is differentiated again, the result is zero, which is easy to verify taking into account that

$$\frac{\partial^2 F}{\partial x_i \partial x_j} = \frac{\partial^2 F}{\partial x_j \partial x_i}\,.$$

The converse statement is also true: if $d\omega^l = 0$, then ω^l is the external differential of some differential forms:

$$\omega^l = d\omega^{l-1}.$$

The expression $\omega^1 = f(x)\,dx$ has been considered as the integrand. It can also be viewed as a linear function of the differential dx. The ω^1 can be assigned a numerical value by substituting the numerical values of x and dx.

Differential forms, which we have introduced as elements of multidimensional integration notation, can also be seen as a function of the point and multilinear functions of the coordinates of the differential vector $(dx_1, dx_2, \ldots, dx_s)$. The numerical value of the differential form can be determined by substituting the numerical values of the coordinates and components of the differential. In this case, it is necessary to rearrange the differentials in descending order of indices, after which the wedge symbol ($\wedge$) can be disregarded when multiplying.

The apparatus of differential forms is an essential element in certain geometric studies, in particular, in the theory of the *symplectic spaces*. However, delving too deeply into mathematical questions is beyond the scope of this discussion.

E.2 *New definition of canonical transformations*

Using external differential forms, the definition (43.2) of the canonical transformation

$$\sum_{i=1}^{s} p_i dq_i - \sum_{i=1}^{s} P_i dQ_i = dF \tag{E.7}$$

can be expressed in a more symmetrical form, by taking from both sides of this equality the external differential and taking into account that the external differential of (dF) equals zero:

$$\sum_{i=1}^{s} dp_i \wedge dq_i = \sum_{i=1}^{s} dP_i \wedge dQ_i \,. \tag{E.8}$$

Of course, by excluding the function $F(q_1, \ldots, Q_s)$ from the definition, the constructive nature of the definition is lost. However, there still exists the option to step back and return from symmetrical definition (8) to constructive (7).

E.3 *Conservation of the phase volume under canonical transformations*

If both sides of the equation (8) are raised to the power of s and the resulting expression simplified, taking into account that only the products of different differentials are nonzero, we obtain:

$$s! \prod_{i=1}^{s} dp_i \wedge dq_i = s! \prod_{i=1}^{s} dP_i \wedge dQ_i \,.$$

Furthermore, considering the equation (5), we can derive:

$$\frac{\partial(Q_1, P_1, \ldots, Q_s, P_s)}{\partial(q_1, p_1, \ldots, q_s, p_s)} = 1 \,. \tag{E.9}$$

This implies that the phase volume is conserved under canonical transformations.

E.4 *Invariance of the Poisson brackets under the canonical transformations*

Let f and g be functions of generalized coordinates and momenta:

$$f = f(q, p), \quad g = g(q, p) \,.$$

We shall now examine the external differential form:

$$df \wedge dg \wedge (\omega)^{s-1} \,, \tag{E.10}$$

where ω is the form,[2] which is included in equality (8). It easy to prove that this expression reduces to

$$(s-1)! \, \{f, g\} \prod_{i=1}^{s} dp_i \wedge dq_i \,, \tag{E.11}$$

where $\{f, g\}$ is the Poisson bracket defined by differentiation with respect to the variables p, q.

[2] Here $(\omega)^{s-1}$ is ω to the power of $s-1$.

The same expression (10) can be written in terms of the canonical variables P, Q. In this case, it only differs from equation (11) in that all variables $p,\ q$ are replaced by $P,\ Q$, including the Poisson brackets. The latter will be calculated by differentiating with respect to new variables: $\{f, g\}_{P,Q}$. Taking into account (8) and (9), this immediately implies invariance of the Poisson brackets under canonical transformations.

Bibliography

[1] L. D. Landau, E. M. Lifshitz (1976) *Mechanics*, Pergamon Press, Oxford.

[2] H. Goldstein, C. Poole, J. Safko (2000) *Classical Mechanics*, Addison-Wesley, Reading, Mass.

[3] G.L. Kotkin, V.G. Serbo (2020) *Exploring Classical Mechanics. A Collection of 350+ Solved Problems for Students, Lecturers and Researchers*, Oxford University Press.

[4] V. I. Arnold (1989) *Mathematical Methods of Classical Mechanics*, Springer-Verlag, Berlin.

[5] N. F. Mott, H. S. W. Massey (1985) *The Theory of Atomic Collisions*, Oxford, at the Clarendon Press.

[6] L. D. Landau, E.M. Lifshitz (1980) *Statistical Physics*, Chap. XIV, Pergamon Press.

[7] N.N. Bogolyubov, D. V. Shirkov (1980) *Introduction to the Theory of Quantized Fields*, John Wiley & Sons.

[8] P. Courant, D. Hilbert (1989) *Methods of Mathematical Physics*, Wiley.

[9] L. D. Landau, E.M. Lifshitz (1984) *Electrodynamics of Continuous Media*, Pergamon Press, Oxford.

[10] L. D. Landau, E. M. Lifshitz (1980) *Classical Theory of Fields*, Butterworth-Heinemann.

[11] V.G. Serbo, V.S. Cherkassky (2017) *Selected Chapters of Analytical Mechanics (Electronic textbook with dynamic interactive illustrations)*, Regular and Chaotic Dynamics, Moscow-Izhevsk.

[12] D. Budker, D. Kimball, D. DeMille (2003) *Atomic Physics. An Exploration Through Problems and Solutions*, Oxford University Press.

[13] E. Fermi (1971) *Scientific papers*, Vol. I, p. 440, Nauka, Moscow.

[14] N. Bloembergen (1965) *Nonlinear optics*, Appendix I, §§ 5,6, W. A. Benjamin, New York.

[15] A. I. Bazh, Ya. B. Zeldovich, A. M. Perelomov (1971) *Scattering, Reactions and Decays in Non-Relativistic Quantum Mechanics*, Nauka, Moscow.

[16] C. Kittel (1968) *Introduction to Solid State Physics*, Wiley, New York.

[17] F. D. Stacey (1969) *Physics of the Earth*, John Wiley & Sons, New York.

Index

Action 33, 149
Adiabatic approximation 166, 168
Allowed band 102, 105
Angle and action variables 160
Anti-oscillator 84, 85
  in magnetic field 88
Aphelion 9

Beam focusing 154
Beats 74

Classical EPR and NMR model 137
Centrifugal
  energy 5, 58
  inertia force 59
Conservation law
  angular momentum 55
  energy 55
  momentum 54
Combination frequencies 111
Constraints
  ideal holonomic 42, 44, 194
  non-holonomic 197, 199
Coordinates
  cyclic 46
  generalized 30
  normal 71
Coriolis force 59
Coupled oscillators 72
Coupled pendulums 72
Covariance of the Lagrangian equations 30, 33
Cyclic coordinates 46

Damped oscillations 79
Damping
  aperiodic 79
  coefficient 79
d'Alembert's principle 44
Dark matter 28
Deviation from the vertical 60
Dipole, motion in the dipole field 52, 129
Drift 39, 40
Dynamic symmetry of the Kepler problem 12, 137
Dynamic chaos 173

Eccentricity 8
Effective potential energy 5, 7, 37
Eigenfrequency 68
Eigenvector 68
Electrical chains 63, 98
Ellipse 9, 18
  focus 9
  major and minor axes 9
Energy flux 102
Equation
  Euler 183, 194
  Hill 200
  Mathieu 120, 201
Euler angles 189

Falling to the centre of the field 16, 21
Finite motion 2
Forbidden band 102, 106
Force of inertia due to uneven rotation 59
Friction 79
Frequency degeneracy 71
Frequency doubling 117

Generalized
  coordinates 30
  momenta 30
Generating function
  for canonical variables 141, 143
  for identity transformation 144
  for time shift 144, 151
Gravity, motion near the surface of the Earth 60
Gyrocompass 187
Gyroscopic forces 82

Homogeneity
  of space 54
  of time 55
Hyperbola 8

Infinite motion 2
Inertia force due to uneven rotation 59
Inertia tensor 178, 179, 180
Integrals of motion 1, 4, 18, 46, 47, 52, 127, 129, 135
Isotropy of space 54

Lagrange points 92
Lagrangian multipliers 198
Laplace vector 11
Light, propagation in medium 127, 130
Localized oscillations 74

Magnetic trap 170
Magnetic field, motion in it 128, 134
Matrix
  of mass 67
  of stiffness 67
Matryoshka 185
Molecules 93, 96
  four-atomic 93
  triatomic 93, 98
Moment of inertia 180
Moving coordinate system 174

Nodes, line 189
Normal
  coordinates 71
  oscillations 68
Nutation 186

Orbit parameter 8
Orbit precession 12, 18, 159
Orthogonality of normal oscillations 70
Oscillation mode 68
Oscillations of linear chains 98, 101, 103
  acoustic 105
  optical 105
Oscillator
  anharmonic 109, 111, 130, 149
  in magnetic field 83, 84
  isotropic 17

Parabola 8
Partial frequency 73
Pendulum 42
  double flat 45
  of Kapitsa 122
Penning trap 89
Perihelion 9
Period of oscillations 2
Period of radial oscillations 5
Perturbation theory 109
Precedence of the equinoxes, 187
Principal axes of inertia 180
Principal moments of inertia 180

Probability in coordinate space 3
Proper rotation 184

Radial motion 4
Rainbow scattering 24
Reduced mass 19
Regular precession 184
Resonance 76, 81
  Fermi 117
  non-linear 113
  parametric 121, 202, 204
Rolling 196
Rotating reference frame 58

Scattering
  at low angles 21, 22
  Rutherford 24
Sectoral velocity 6
System of the Earth–Moon 14, 188
Symmetry properties of normal oscillations 93

Tidal force 188
Theorem
  Huygens–Steiner 180
  Larmor 59
  Liouville 152, 211
  Noether 51, 151
  Poisson 135
  virial 26
Top
  asymmetrical 181, 183
  fast 186
  spherical 180, 183
  symmetrical 181, 183
  with a fixed fulcrum 186, 187
Transformations
  Galilean 55
  gauge 35
  Lorentz 207
  similarity 52
Translational reference frame 57
Turning point 2, 5

Variation 192
Variational derivative 194

Wave vector 101
Waves
  standing 102, 105
  travelling 99
Watt's regulator 46